Breeds of Empire

NORDIC INSTITUTE OF ASIAN STUDIES
NIAS Studies in Asian Topics

16 Leadership on Java *Hans Antlöv and Sven Cederroth (eds)*

17 Vietnam in a Changing World *Irene Nørlund, Carolyn Gates and Vu Cao Dam (eds)*

18 Asian Perceptions of Nature *Ole Bruun and Arne Kalland (eds)*

19 Imperial Policy and Southeast Asian Nationalism *Hans Antlöv and Stein Tønnesson* (eds)

20 The Village Concept in the Transformation of Rural Southeast Asia *Mason C Hoadley and Christer Gunnarsson (eds)*

21 Identity in Asian Literature *Lisbeth Littrup* (ed.)

22 Mongolia in Transition *Ole Bruun and Ole Odgaard* (eds)

23 Asian Forms of the Nation *Stein Tønnesson and Hans Antlöv (eds)*

24 The Eternal Storyteller *Vibeke Børdahl (ed.)*

25 Japanese Influences and Presences in Asia *Marie Söderberg and Ian Reader (eds)*

26 Muslim Diversity *Leif Manger (ed.)*

27 Women and Households in Indonesia *Juliette Koning, Marleen Nolten, Janet Rodenburg and Ratna Saptari (eds)*

28 The House in Southeast Asia *Stephen Sparkes and Signe Howell (eds)*

29 Rethinking Development in East Asia *Pietro P. Masina (ed.)*

30 Coming of Age in South and Southeast Asia *Lenore Manderson and Pranee Liamputtong (eds)*

31 Imperial Japan and National Identities in Asia, 1895–1945 *Li Narangoa and Robert Cribb (eds)*

32 Contesting Visions of the Lao Past *Christopher Goscha and Søren Ivarsson (eds)*

33 Reaching for the Dream *Melanie Beresford and Tran Ngoc Angie (eds)*

34 Mongols from Country to City *Ole Bruun and Li Naragoa (eds)*

35 Four Masters of Chinese Storytelling *Vibeke Børdahl, Fei Li and Huang Ying (eds)*

36 The Power of Ideas *Claudia Derichs and Thomas Heberer (eds)*

37 Beyond the Green Myth *Peter Sercombe and Bernard Sellato (eds)*

38 Kinship and Food in South-East Asia *Monica Janowski and Fiona Kerlogue (eds)*

39 Exploring Ethnic Diversity in Burma *Mikael Gravers (ed.)*

40 Politics, Culture and Self: East Asian and North European Attitudes *Geir Helgesen and Søren Risbjerg Thomsen (eds)*

41 Beyond Chinatown *Mette Thunø (ed.)*

42 Breeds of Empire: The 'Invention' of the Horse in Southeast Asia and Southern Africa 1500–1950 *Greg Bankoff and Sandra Swart*

Breeds of Empire

The 'Invention' of the Horse in Southeast Asia and Southern Africa 1500–1950

Greg Bankoff and Sandra Swart

with Peter Boomgaard, William Clarence-Smith,
Bernice de Jong Boers and Dhiravat na Pombejra

 niasPRESS

Breeds of Empire: The 'Invention' of the Horse in Southeast Asia and
Southern Africa 1500-1950

By Greg Bankoff and Sandra Swart, with Peter Boomgaard,
William Clarence-Smith, Bernice de Jong Boers and Dhiravat na Pombejra

NIAS – Nordic Institute of Asian Studies
Studies in Asian Topics, no. 42

First published in 2007 by NIAS Press
Nordic Institute of Asian Studies (NIAS)
Leifsgade 33, 2300 Copenhagen S, Denmark
E-mail: books@nias.ku.dk
Online: http://www.niaspress.dk/

British Library Cataloguing in Publication Data

Bankoff, Greg
  Breeds of empire : the 'invention' of the horse in Southeast Asia and
  Southern Africa, 1500-1950. - (NIAS studies in Asian topics ; 42)
  1. Horses - Thailand - History 2. Horses - Philippines - History 3.
  Horses - Africa, Southern - History 4. Horses - Indonesia - History 5.
  Horses - Breeding - History 6. Southeast Asia - Economic conditions
  7. Southeast Asia - Foreign economic relations 8. Africa, Southern -
  Economic conditions 9. Africa, Southern - Foreign economic relations
  I. Title II. Swart, Sandra
  636.1'00959

ISBN: 978-87-7694-014-0 (hardback)
ISBN: 978-87-7694-021-8 (paperback)
ISSN: 0142-6028

Typesetting by Donald B. Wagner

Produced by SRM Production Services Sdn Bhd
and printed in Malaysia

Contents

Contributors   vii

1   Breeds of Empire and the 'Invention' of the Horse   1
    Greg Bankoff and Sandra Swart

Part One

2   Southeast Asia and Southern Africa in the Maritime Horse Trade of
    the Indian Ocean, c. 1800–1914   21
    William Gervase Clarence-Smith

3   Horse Breeding, Long-distance Horse Trading and Royal Courts in
    Indonesian History, 1500–1900   33
    Peter Boomgaard

4   The 'Arab' of the Indonesian Archipelago: Famed Horse Breeds of
    Sumbawa   51
    Bernice de Jong Boers

5   Javanese Horses for the Court of Ayutthaya   65
    Dhiravat na Pombejra

Part Two

6   Colonising New Lands: Horses in the Philippines   85
    Greg Bankoff

7   Adapting to a New Environment: The Philippine Horse   105
    Greg Bankoff

8   Riding High – Horses, Power and Settler Society in Southern
    Africa, c. 1654–1840   123
    Sandra Swart

9   The 'Ox That Deceives': The Meanings of the 'Basotho Pony' in
    Southern Africa   141
    Sandra Swart

Epilogue

10 'Together yet Apart': Towards a *Horse*-story   153
    Greg Bankoff and Sandra Swart

    Notes   155
    Bibliography   219
    Index   249

Maps

3.1 Horse Breeding and Trade in Indonesian History.   49
4.1 The Island of Sumbawa   52
4.2 The Indonesian Archipelago   53
5.1 Ayutthaya and Java   66
5.2 Siam   67
6.1 The Philippines   88
9.1 Southern Africa   142

Figures

6.1 Variations in Equine Population on Luzon, 1886-1918   96
7.1 Comparison of Equine Height to Numbers, Philippines, 1935 106
8.1 Growth of the Horse Population in the Cape, 1652-1702   130
8.2 A View of Lord Charles Somerset   138
9.1 Photograph from the Remount Commission, Basotho Pony, c. 1902 146

Tables

2.1 Horse Exports from Southern Africa to all Destinations,
     1827-1918   25
2.2 Mauritian Imports of Equids 1833-1913   27
2.3 Horse Exports from the Lesser Sunda Islands, 1880-1918   30
3.1 Number of Horses per 1,000 People, Java, 1820-1900   44
6.1 Number of Conveyances Registered in Various Provinces,
     Philippines, 1889   87
6.2 Horses Employed by Activity, Philippines, 1889   90
6.3 Equine Population Changes, Philippines, 1886-1903   93
6.4 Causes of Mortality Among Equine Populations by Region,
     Philippines, 1918   99
6.5 Domesticated Equine Population Densities per Square Kilometre,
     Philippines, 1886   101

Front Cover

Lace. Photographer Graham Walker.
Three horse riders in ceremonial clothing on the occasion of the wedding between the Sultan of Sumbawa and the daughter of the Sultan of Bima. Photographer/Copyright Royal Netherlands Institute of Southeast Asian and Caribbean Studies (KITLV), Leiden, no. 29.508. Date: 1931.

Back Cover

Miss Anna Ide, mounted, Baguio, Benguet. Photographer/Copyright Museum of Anthropology, University of Michigan, Dean C. Worcester Photographic Collection, 1890-1913, no. 58M064. Date: 1904.

Contributors

Greg Bankoff is a social and environmental historian of Southeast Asia and the Pacific. In particular, he writes on environmental–society interactions with respect to natural hazards, resources, human–animal relations, and issues of social equity and labour. He is professor of modern history in the Department of History, University of Hull. Among his publications are *Crime, Society and the State in the Nineteenth Century Philippines* (Manila: Ateneo de Manila University Press, 1996) and *Cultures of Disaster: Society and Natural Hazard in the Philippines* (London: RoutledgeCurzon, 2003). He is also co-editor of *Mapping Vulnerability: Disasters, Development and People* (with Georg Frerks and Dorothea Hilhorst, London: Earthscan, 2004).

Peter Boomgaard was trained as an economic and social historian and obtained his PhD from the Vrije Universiteit, Amsterdam, in 1987. He is Senior Researcher at the Royal Netherlands Institute of Southeast Asian and Caribbean Studies (KITLV), Leiden, and Professor of Environmental History of Southeast Asia at the University of Amsterdam. Among his publications are *Children of the Colonial State: Population Growth and Economic Development in Java, 1795–1880* (Amsterdam: Free University Press, 1989), *Frontiers of Fear: Tigers and People in the Malay World, 1600–1950* (New Haven/London: Yale University Press, 2001), and *Southeast Asia: An Environmental History* (Santa Barbara: ABC-CLIO, 2007).

William G. Clarence-Smith is Professor of the Economic History of Asia and Africa at the School of Oriental and African Studies, University of London. He has worked extensively on commodities and traders. In his first book, *Slaves, Peasants and Capitalists in Southern Angola, 1840–1926* (Cambridge University Press, 1979) he touched briefly on the horse trade across the Kalahari. He is currently researching the history of equids and elephants in Southeast Asia and its borderlands, c1760-c1960. His most recent book is *Islam and the Abolition of Slavery*, London: Hurst, 2006.

Dhiravat na Pombejra is the author of several articles on Thai history, including "Ayutthaya at the end of the seventeenth century: was there a shift to isolation?" in Anthony Reid (ed.), *Southeast Asia in the Early Modern*

Era: Trade, Power, and Belief (Ithaca: Cornell University, 1993) and "VOC employees and their relationships with Mon and Siamese women: a case study of Osoet Pegua" in Barbara Watson Andaya (ed.), *Other Pasts: Women, Gender and History in Early Modern Southeast Asia* (Honolulu: Center for Southeast Asian Studies, University of Hawai'i at Manoa, 2000). He is co-editor, with Remco Raben, of *In the King's Trail* (Bangkok: Royal Netherlands Embassy, 1997). Formerly a Lecturer in History at Chulalongkorn University, Bangkok, he continues to teach and to do research, mainly on the history of seventeenth and eighteenth century Siam.

Bernice de Jong Boers was trained as an anthropologist at Utrecht University and was affiliated as a PhD student at the Royal Netherlands Institute of Southeast Asian and Caribbean Studies (KITLV) at Leiden from 1993 to 1997. Her research concentrated on the environmental history of the island of Sumbawa. She is currently employed as a programme officer at the Social Sciences Research Council of the Netherlands Organisation for Scientific Research (NWO).

Sandra Swart is an environmental and social historian of southern Africa. She received both a DPhil in Modern History and MSc in Environmental Change from the University of Oxford. She is a Senior Lecturer at the University of Stellenbosch. She has published on themes as various as Afrikaner identity, animals in history and social rebellion in, amongst others, the *Journal of Southern African Studies* and *Journal of African History*. She is the co-editor, with Lance van Sittert, of *Canis Africanis: A Dog History of Southern Africa* and she is currently writing the history of horses in southern Africa.

1

Breeds of Empire and the 'Invention' of the Horse

Greg Bankoff and Sandra Swart

An animal is not simply 'invented' as such, nor is breeding simply the outcome of human endeavour. A breed is as much the product of the subtle influence of environmental factors and the animal's innate adaptive abilities as it is of purposeful human manipulation. Together they shape the animal's physiological form and give rise to its identifiable characteristics as a specific strain or variety that can then be breed for inheritance. The effects of these three factors are evident in the horse (*Equus caballus*), an animal that has been the object of concerted human interest and concern over millennia.[1] The horse was domesticated around five thousand years ago in the steppes of central Asia and its use spread into China, India, western Asia, Europe and Africa sometime during the second millennium BCE.[2] The close association between the horse on the one hand and state, conquest and empire on the other is a recurring motif in the animal's history, dramatically displayed in the Spanish conquest of the Americas in the fifteenth and sixteenth centuries but no less a feature as well of the European conquest and colonisation of southern Africa in the seventeenth century and Australasia in the eighteen and nineteenth centuries.[3]

Empires are usually seen as exclusively human endeavours, the affairs of men (and sometimes women), while horses are viewed as the product of the environment and of natural selection as well as human breeding programmes. While modern historiography is beginning to explore the importance of animals and to discover that they may even have their own histories independent of their utilitarian and subservient roles in human affairs, little attention has so far been paid to the effect of empire and breeding.[4] The horse was very much an imperial agent and constituted an important part of Alfred Crosby's 'portmanteau biota' that followed European colonies and settlement across the globe from the late fifteenth century.[5] Not only soldiers and settlers but also equine genes and phenotypes were disseminated along imperial networks of trade and diplomacy

1

as colonial governments and the societies they created sought to introduce or improve the horse for military, infrastructural or agricultural purposes. On the one hand, colonists and European settlers at the fringe of empire created new breeds of horses to suit their own needs through importing horses and cross-breeding them with local ones or with one another. On the other hand, indigenous peoples, less concerned by questions of pedigree and size, pursued their own designs. The resulting new colonial breeds differed markedly from those of the metropole and even, in some instances, came to be identified with particular colonial or indigenous cultures. In the process, new breeds – 'breeds of empire' – were 'invented' from which the modern horse in many countries derives its origin.

Animals in History

Animals, as Claude Lévi-Strauss observed, are 'good to think' with. Because of their central place within human society and the powerful emotions they evoke, they may afford scholars a window into wider socio-historical processes.[6] This collection of essays reassesses the variety of ways in which animals – in this case, horses – were utilised and conceived of on the periphery of empire. As Crosby noted 'man' came not to the colonised world 'as an individual immigrant but as part of a grunting, lowing, neighing, crowing, chirping, snarling, buzzing, self-replicating and world-altering avalanche'.[7] Just as they had done in Europe, Asia and North Africa, horses not only provided power and transportation to indigenous people and colonisers but also altered their new environments in various ways.[8] This compilation may be read as part of what has been termed the 'animal moment' in the social sciences.[9] Recent historiography is beginning to give greater emphasis to the importance of animal-centred research and, as Harriet Ritvo has observed, no longer is the mention of an animal-related topic likely to provoke 'surprise and amusement' as was the case twenty years ago.[10]

'Animal studies' is now a growing academic field, and wide-ranging in disciplinary terms: including, for example, anthropologies of human–animal interactions, animal geographies, the position of animals in the construction of identity, and animals in popular culture.[11] This new direction, as Chris Philo has commented, was inspired or at least encouraged by the encounter with ideas from social theory, cultural studies, post-colonial studies, and psychology.[12] The 'animal turn' explores the spaces which animals occupy in human society and the manner in which animal and human lives intersect, showing how diverse human groups construct a range of identities for themselves (and for others) in terms of animals. It is argued that attention should turn to the cultural meaning of non-humans in the histories, anthropologies, sociologies and geographies – the stories – of everyday life.[13] Rather than conceptualising animals as simply elements of ecological, agricultural or bio-geographical arrangements, they can be understood in the context of their role – both passive and active – in human cultures. What is emphasised here is the consideration of an extended intellectual territory in which animals are more than either 'resources' or 'units of production'. This has initiated the analysis of the anthropogenic use of animals as something other than either 'natural' or unproblematic.[14]

Perhaps, as Jennifer Wolch and Jody Emel contend, it is the very centrality of animals to human lives that has previously rendered them invisible, at least invisible to scholars intent on mainstream history.[15] Of course, animals *per se* can hardly be described as uncommon historical subjects.[16] Their remains – both corporeal and pictorial – have provided sources for societies that left no written evidence.[17] Moreover, as Ritvo has observed, they have been studied by economic historians (interested in agricultural sectors); by social historians (intrigued, for example, by animal-related social institutions)[18] and by biographers (who have written of people 'involved with animals' like hunters, and, indeed, even some animals – like the race horses, *Man o' War* and *Seabiscuit*).

Recent years have witnessed a transformation in the attitude of historians toward the study of animals. The first change is quantitative: Animals have simply drawn greater academic interest. Arguably this is because changes in socio-political ideas are usually echoed by the themes explored by historians. Thus, as Ritvo has contended, just as the field of labour history followed the rise of the labour movement and the sub-disciplines of women's history and African–American history followed the women's movement and the civil rights movement, in the last 25 years, animal-related causes have gained increasing support in the west. To follow Ritvo, animals can be seen thus as the latest beneficiaries of a 'democratising tendency' within historical studies.[19] Moreover, the development of environmental history has encouraged historians toward studying animals: many of the issues at the crossroads of academic studies of the environment and environmental politics have an animal component, which increasingly renders historical studies such as this, which take an inter-disciplinary approach to compensate for a lack of written evidence, important. Wolch and Emel contend that human practices now threaten animal worlds – indeed, the global environment – to such an extent that humans have now both an 'intellectual responsibility' and 'ethical duty' to consider animals closely.[20] Perhaps additionally, the ethological observations of animals as closer to humans than we have acknowledged so far, the growing rejections of the universality of the nature–culture distinction, and the human-ecological emphasis on the influence of animals on human culture, all lead towards a gradual rejection of the nature–culture distinction that has been a central part of the distinction between social and natural sciences. Other theorists argue that animals were never part of the modernist project – except, arguably, as commodities – and in the post-modern moment, particularly with the rise of the animal rights movement, there is increasing attention from the left-leaning scholars and activists.[21] Steve Baker and others have argued that the post-modern project of de-centring the human subject creates space to place animals within the axis of scholarly scrutiny so that animals will loom larger in historical accounts.[22]

The second change is qualitative. Formerly human–animal interactions have been studied by historians (and other social scientists) with the principal purpose of an improved understanding of humans' relationships with other humans. Significance and values ascribed to animals, systems of categor-ising them and ways of utilising them (as food, commodities, symbols or 'stand-in humans') have been addressed by researchers, particularly anthropologists.[23] Animals exist in the academic literature, but there has been a general tendency to regard them as raw material for human action (especially economic) and human thoughts, and a – perhaps concomitant – propensity to represent them as passive objects.[24] The active role of animals in their interactions with humans has so far not been addressed by historians, with the nature–culture dichotomy still strong in their imagin-ations. Peter Boomgaard has proposed the possibility of writing history that incorporates the perspective of animals.[25] As more historians have chosen to work on animal-related topics, such topics increasingly have been inte-grated into the disciplinary mainstream. It reflects two convergent trends, as Ritvo has noted. One is the readiness of historians who work on other topics to recognise the historical significance of animals. The other is the tendency of historians who work on animal-related topics to present them as part of the general history (of a particular period and location) rather than ghettoising them in peripheral sub-fields.[26]

Part of this 'turn' looks at the role of animals in identity construction. As Wolch and Emel have contended, it remains largely un- or at least under-discussed in the discourse around identities how animals play a role in the construction of that identity.[27] Roy Willis finds in the way that people relate to animals 'a key to read off certain otherwise inaccessible information about the way human beings conceived of themselves and the ultimate meaning of their own lives'.[28] Paul Shepard and others have initiated the discussion on a broad level, arguing that humans as a species define themselves in relation to animals.[29] In this collection, however, as in the case studies discussed below, we focus on the subtle, idiosyncratic identities that are either adopted or imposed by groups based on, for example, geographic location, race, ethnicity or gender. Philo has delineated the appearance of a particular urban identity in Victorian Britain associated with standards of civilised behaviour that stood in distinction to rural areas and how live meat markets and abattoirs breached these budding standards and so were banished from the city.[30] Kay Anderson, for example, has demonstrated how zoo practices served to naturalise the imposition of colonial rule and to consolidate and legitimise Australian colonial hegemonic identity.[31] Jody Emel has argued that a particular form of masculine identity was associated with wolf eradication attempts in the American West in the late nineteenth century, with dominant representations of masculinity counterposed to so-called lupine traits like 'cowardice' and 'lack of mercy'.[32] Various theorists have argued that individuals and communities tend to attribute their own dreams and desires to animals, including those 'natural impulses' that they perhaps most dread in themselves.[33]

Some recent studies point to how particular human societies wished to exclude particular kinds of animals from their homes (because they were regarded as 'useless', 'wild' or 'unclean'). Other studies, like this one, focus not on exclusions, but inclusions – where some animals were incorporated (as 'useful', 'tame', 'noble' or 'charismatic'). These studies, for example, trace the keeping of domestic animals, like the horse, within the domestic, military and commercial milieu. This departure focuses on the changing symbolic dimensions of the encounter between humans and animals, coupled with material societal changes. Such codings and transformations become closely entangled in the identity politics of human groups, with animal images and metaphors deployed to reflect human societal strata and vice versa.

Whether the animal 'turn' is manifested in mainstream historical work or featured in the inter-disciplinary domain of 'animal studies' makes little difference to either the methodology or the simple fact that, as Ritvo notes, historical research provides the context for more exclusively interpretive scholarship. It is in this spirit that this collection seeks to contribute to the ongoing historical project of understanding our shared past. Materialists believe nature to be a basic decisive factor for culture.[34] Structuralists or symbolists have often taken for granted Claude Lévi-Strauss's postulate that all cultures set boundaries between nature and culture, and use the opposition as an analytical tool to understand myths, rituals, classification systems and the like. Lévi Strauss attempted, for instance, to show how differences between species of animals can be good to think with for humans, to illustrate differences between human groups. For Lévi-Strauss, animal species, with their many observable differences and habits, offered a way of naturalising social classifications and distinctions.[35]

The Animal 'Other'

Animals, then, have increasing begun to figure in 'our' histories. As Ritvo argues they form part of our epic tales about the early movement of hunters and gathers, part of the grand narrative of domestication and agricultural transformation, have represented nature in religious and scientific thought, and have figured allegorically and increasingly as moral reminders in literature.[36] But in a very real sense, they are still depicted and talked about as if they were some animal 'other', or, as Erica Fudge so poignantly expresses it, as 'absent yet present' in history.[37] Edward Said's 'other', of course, was the Orient, an intellectual framework he claimed was constructed by western scholars and travellers to come to terms with and even manage the East, 'dealing with it by making statements about it, authorising views of it, describing it, by teaching it, settling it, ruling over it'. The idea of hegemony is central to Said's analysis, a hegemony that is implied by the very structures of the disciplines, expressed in the language employed and assumed in the histories written.[38] It is these same western 'men' (and increasingly women, itself an interesting development) who are now mainly writing about animals. Moreover, they are principally writing the histories of 'western' animals. There is not only an animal 'other' but it is still largely an 'orientalist' one.

The historical 'othering' of animals starts with denying non-human species a separate sense of independent history. If animals appear in the narrative, they do so mainly as adjuncts or backdrops to people and their activities. While such recognition provides some historical record, it rarely says anything more significant *about* them. The implicit assumption is that only humans have narratives and therefore presumably histories. Animals are reduced to a sense of time that can only be expressed in terms of genetic inheritance and lack any ability to mark its passage. In particular, the want of verbal or written sources and above all their inability to express consciousness deprives them from any participation in history.[39] Attempts to endow animals with a degree of consciousness, a state of awareness that if accepted might challenge such an attitude, are branded anthropomorphic and, until recently, dismissed out of hand.[40] Yet the relationship of history, narrative and anthropomorphism is central to the construction of a more animal-centric historiography. Narrative is generally considered the defining feature of historiography, the notion that there is a continuum of explanations that connects one event to another so that a plot is developed.[41] While this confers an inevitable teleological sense of certainty about the narrative, the factors that primarily decide the outcome are the parameters, determining the start and end points of the plot. Within these, the essential element is agency: that the chosen protagonist has an awareness of self and thereby the potential to influence situations for his or her own purpose. Without such a sense of consciousness, the subject remains passive, only so much matter to be acted upon or transformed by others. The argument that denies animals any sense of self-awareness is therefore central to the role they are usually credited with in history and to the 'othering' with which they are tarnished.

The modern debate about animal intellect traces its origin to the writings of Charles Darwin.[42] Early animal psychologists such as George Romanes certainly credited them with intelligence and attempted to clarify its extent and nature through classifying the many anecdotal descriptions given by mid nineteenth century naturalists.[43] However, the development of experimental psychology associated particularly with the laboratory studies of Edward Lee Thorndike adversely influenced the manner in which animals came to be regarded during the early decades of the new century.[44] They were transformed from objects with albeit diminished natural intelligence to subjects of psychological study whose behaviour could be elicited by the appropriate stimuli.[45] The resultant enthusiasm for this 'more scientific' approach to psychology has become the dominant paradigm since the 1920s. All behaviour, no matter how complex, was explained in terms of learnt responses with no reference to subjective reflection: 'animal behaviour was defined as what could be observed and measured in the laboratory' and animals were denied all sense of a cultural life.[46]

Any attempt to credit animals with feelings, motivations and thoughts were subsequently seen as anthropomorphic, ascribing human mental experiences to animals. Behaviourists still maintain that the existence of consciousness in animals, that is the existence of an 'inner eye' or self-monitoring system, lies beyond the scope of rigorous scientific proof.[47] Only more recently has this certainty been increasingly assailed on the grounds that it does not conform to Darwinian orthodoxy: if consciousness has evolved, it must confer some advantage and so make a detectable difference in how an organism behaves. Marian Dawkins argues that consciousness and emotions are closely linked and that the intensity of both is heightened at moments of acquiring new skills or dealing with novel or unexpected situations. She maintains that there is sufficient evidence to conclude that mammals and even birds experience forms of consciousness that are at least somewhat similar to humans and that: 'we have to be prepared to go beyond the narrow-minded, rather arrogant anthropomorphism that sees human conscious experiences as the only or even the ultimate way of experiencing the world and make ourselves open to the much more exciting prospect of discovering completely new realms of awareness'.[48]

This general 'othering' of animals bears more than a passing similarity to that meted out to other actors whose historical import has been downplayed or largely ignored until recently. In particular, the depiction of women as largely passive victims completely outside of time, 'hidden' and 'invisible' agents ('absent yet present') provides many parallels to the current treatment of animals' historical role.[49] Just as the distinction between 'nature' and 'culture' was employed as a discursive device to devalue women and their bodies in comparison to men and their activities, at least from the eighteenth century, so the same association is still often used to deprive animals of all sense of any culture, volition or even consciousness.[50] The implied relational hierarchy of inferiority to superiority, according to Gisela Blok may not be simply sexist but may even be specifically western, raising some intriguing questions as to the cultural ontology of how non-humans are generally related to humans.[51]

It has been claimed that women's history revitalised the foundations of historiography by problematising three of the basic concerns of historical thought: periodisation, categories of social analysis, and theories of social change.[52] How much would modern historiography have to be recast if the human–animal relationship was made a significant determining factor? Certainly the replacement of animal power by that of the mechanical engine would have to be considered a critical historical turning point. But this was a gradual process that partly coincided with the later developments of the Industrial Revolution and so would not necessarily require a complete revision of the current periodisation, though the justification between stages might have to be signified more clearly. Social analysis would be more radically altered if the relational importance between human and animal emerged as a major category of social thought in addition to race, ethnicity, class and gender. Here, too, the whole concept of what was meant by 'animal' might need reappraisal in the light of Donna Haraway's accusations that present definitions are nothing more than western cultural constructions, what she calls 'simian orientalism'.[53] Moreover, any theory of social change that attempts to incorporate the relationships between human and animal would have to consider how their relative positions have been affected by both changes in science, technology and production as well as adaptation, mutation and population. Again, such a comparison may entail a re-evaluation of whether what is generally considered historical fact is implicitly defined in terms of purely human action.

All these arguments apply even more forcibly when it comes to considering animals in a non-western setting. Even the more recent scholarship that fully accords animals both pasts and histories is, with few exceptions, still exclusively Eurocentric or neo-Eurocentric, about animals in Europe or in its off-shoot settler societies.[54] Animals in the rest of the world are effectively unknown except as far as they constitute a source of wild and exotic species to take on tour for spectacle, to stock the cages of western zoos with or to provide hapless victims to the rifles (or butts) of imperial sportsmen.[55] The focus was on the exotic or the epitome of wilderness – the elephant, rhinoceros, lion or tiger. There are few exceptions to this bias, most notably Richard Bulliet's work on the camel in Islamic society, Robin Law on the role of the horse in West Africa, Peter Boomgaard's study of tigers and indigenous people in the Malay world, L. Van Sittert and Sandra Swart on the dog in South Africa, and John Knight's edited volume on cultural perspectives of wildlife in Asia.[56] While the significant role of animals on a global scale is acknowledged in the works of Alfred Crosby, where the transfer of species between Europe and the Americas is viewed as part of a world-altering avalanche, they still largely remain the dumb instruments of ecological change rather than the focus of separate historical inquiry.[57]

Yet even as unwitting agents, other species apart from humans transform a landscape. In some cases, the impact of a single species or order of species can create an environment that better accommodates their own nutritional requirements. Plains and grasslands such as the American Mid-west or the Serengeti in Southeast Africa are not naturally largely treeless expanses but are maintained that way through the innumerable activities of the animals that browse upon them often in huge numbers.[58] The relative absence of trees, in turn, affects the climate leading to further desiccation of the landscape. Such environmental effects can be quite dramatic and some have happened in comparatively recent historical times so that their impact can be measured. Often, these animals may not be native to that landscape but have been introduced through some external agency that is often human but not necessarily so. At other times, human agency may not even be a factor at all or may only be incidental, as in the case of the brown rat (*Mus decumanus*).[59] The introduced species may then explode upon an ecosystem that is or has been left evolutionarily unprepared for such an onslaught.

As Alfred Crosby argues, the Spaniards did not come alone to the New World but brought with them domesticated and undomesticated animals, the grains to feed both, the weeds that came unwittingly in their seed-stock, and even their own specialised varieties of pathogens.[60] Where this introduction involved ungulates (herbivores with hard horny hooves), the animals responded to an excess of available food over the amount required to simply replace their numbers by increasing exponentially until they exceeded the carrying capacity of the ecosystem to support them. This is termed an *ungulate irruption* and has been historically documented in the case of sheep in Mexico by Elinor Melville.[61] Ungulate numbers continue to increase, reaching their greatest densities just prior to the exhaustion of the original standing crop of native vegetation. As the latter reaches its lowest density, the migrant animal population crashes so allowing time and space for the vegetation to recover. After this, animal and plant reach a form of mutual accommodation, a condition termed *irruptive oscillation* that continues to act as a check on both their populations.[62] Though, over time, agricultural progress has overcome these natural ecological ceilings, an ecosystem is irreversibly altered in the process.

Ungulate irruption refers to a universal phenomenon. While interest has been directed towards the introduction of animals into the Neo-Europes, little attention has been given to their effect on more tropical environments. Away from the temperate latitudes, the introduction of Old World ungulates did not have such an obviously devastating impact on the environment but it nevertheless was a significant factor of ecosystem change. Much more neglected, however, is the subsequent transformation of the introduced species itself; how the animal changes the environment and how the environment changes the animal in a process of mutual accommodation. Humans may have a direct or indirect hand in this transformation or their role may be that of largely passive bystanders in the ongoing dynamics. Animals are not only the historiographical 'other', frequently exoticised if mentioned at all, but those outside the greater western or neo-western world are largely invisible and have been generally ignored by historians.

'Inventing' the Horse: Southeast Asia and Southern Africa

The horse in our story has been 'invented' in at least two senses: one, it has been created in the minds of people as a symbolic or representational construction; and two, it has been literally morphologically refashioned by anthropogenic intervention – with the two categories frequently impacting on one another. There is a measure of methodological disagreement on the first point. Historians like Ritvo and ourselves are largely concerned with the 'real animal' (with interpretations produced by those who deal with 'real animals' and records to do with breeding, veterinary medicine, agriculture, and natural history).[63] John Berger and others, however, have tried not to draw that distinction, insisting that he speaks of nothing more real than human imaginings.[64] However, any symbolic or representational use of the animal must receive the same critical attention as the real beast.[65]

Good Breeding

The term 'breed' is notoriously hard to define.[66] As mentioned earlier, a breed is as much the product of persistent human manipulation as it is of environmental influence and the creature's innate adaptive abilities. A 'breed' may be understood as animals that, through selection and reproduction, have come to resemble one another and pass their traits uniformly to their offspring. A breed is smoothly defined as a Mendelian population in equilibrium differentiated from other breeds by genetic composition. Or, to put it another way, a breed is a population that complies with ancestry. So a 'purebred' animal belongs to an identifiable breed complying with prescribed traits – origin, appearance, and minimum breed standards. As J. Lush contends in *The Genetics of Populations* the term, therefore, is both elusive and subjective.[67]

'Animal breeding' transformed the corporeal bodies of its subjects. Clutton-Brock and others have demonstrated that domestication produced similar changes in a variety of species: change in size (for example, cows and sheep became smaller; horses larger); increased diversity in traits like coat colour; and the retention of neotenous characteristics (both morphological and behavioural) into maturity.[68] Jared Diamond and others have postulated that people originally selected animals for tractability and for distinctiveness, which made it simpler both to control the creatures and to distinguish between them. Once domesticated populations were distinguished from their wild relatives, however, humans began to breed for more specialised qualities (like greater speed in horses and larger milk yield in cows).[69]

Modern breeders often aver that their preferred type of horse (dog, cow or pig) possesses an ancient pedigree. Although it is evident that divergent varieties existed in former times, it is hard to make clear-cut connections from them to particular modern types. (Of course, as Ritvo observes, every living animal has ancient ancestors, just as every living human does, but in 'both cases the problem is to figure out who they might be'.) By the mid-1800s, breeding by anthropogenic selection had become common practice and the ambition of breeders was initially to effect measurable change. The first animals for which there were public breeding records were the thoroughbred horse and the greyhound – bred for the easily quantifiable trait of speed. The leading breeder of the latter, Robert Bakewell (1725–1795), purposefully concealed the ancestry of his prized animals: their quality was guaranteed by *his* name rather than those of his animals.[70] In fact, Bakewell learnt much from the selective breeding used to create the thoroughbred race horse.[71] Until that time, animals of mixed gender associated freely, permitting uncontrolled haphazard breeding. Bakewell rigorously separated the sexes, and controlled mating. Moreover, and most importantly, by inbreeding his stock he fixed and amplified sought-after characteristics. He utilised a basic system of progeny testing: leasing sires of known pedigree to nearby farmers, while they supplied Bakewell with records on any brood produced. Bakewell bred the 'best to the best', regardless of their kinship and disregarding the stigma. This breaking of taboo meant inbreeding could fix a desired standard type, eventually leading to the first 'true breeding lines' that produced homogeneous progeny.

Human–animal relations, however, underwent a change between the seventeenth and nineteenth centuries as 'Nature' *per se* was subordinated to human manipulation.[72] Donna Landry has noted that genetic potential was encoded as 'blood' from the seventeenth century, representing ill-defined but potent and innate qualities handed down from generation to generation: homologous with human aristocracy.[73] This notion was refuted in the eighteenth century as scientific investigation into anatomy and biological mechanics changed the human understanding of the bodies and thus the pedigrees of horses.[74]

Another telling case study from the imperial metropole, even before Bakewell's successes, was the way in which anglophone horse breeders experimented in selective breeding, generating what came to be called the 'English Thoroughbred'.[75] These progeny of Eastern imports were 'naturalised' as English subjects within just one generation, ready for re-export as 'English' produce.[76] On a practical level, Joan Thirsk has shown that environmental conditions were perhaps as vital as genetic background for the morphological changes from 'Oriental stock' to Thoroughbred horse. Nutrition – pasturage and cereals that had themselves been improved by experimentation – lent substance to the imported desert horses.[77] On a more symbolic level, the 'Englishness' of the Thoroughbred had its origins not in nature at all but rather in nurture as Landry observes. Landry argues that, paralleling John Locke's argument for colonisation (that labour invested in land gave one the right to claim that land), the new producers of the Thoroughbred horse claimed 'ownership' of their horse's 'national identity'. This is a pattern repeated elsewhere both in the metropole and in the periphery of empire.[78] As information on genetics became common public knowledge, backyard breeders and small studs proliferated, and 'breed standards' became a matter of popular debate.

Southeast Asia

Research on breeds of horses in Southeast Asia largely reflects the same limitations as other animals in the non-western world, only compounded by the absence of records and scholarship.[79] The archaeological remains of horses in mainland Southeast Asia, which are thought to have reached there via Yunnan from China, date from around the end of the third century CE. After the Mongol invasions of the thirteenth century, the use of horses became more widespread especially in warfare.[80] The date of their arrival in the Indonesian archipelago is uncertain though the animal appears to have been introduced sometime before the ninth century CE, though total numbers remained low before 1200.[81] Horses did not reach the Philippine islands until introduced by the Spaniards after 1565, though animals from Indonesia were probably already present in the southern island of Mindanao and possibly in the Sulu archipelago.[82] Apart from the Philippines, horses were closely associated with courts in maritime Southeast Asia, especially those with upland valleys, so that most of the known breeds have royal pedigrees.[83] The animals were generally small, more the size of ponies, though recognised for their endurance and stamina.[84] They were much sought after for pomp and ceremonial purposes, though the extent of their deployment in warfare (unlike on the mainland) is less clear. As horses became more numerous and less a status symbol in the eighteenth century, they were increasingly used for more mundane work such as drayage and cartage, though their employment in agriculture remained limited due to physiognomies unsuited for work in rice paddies.

The advent of the colonial state during the sixteenth and seventeenth centuries in the form of the Dutch United East Indies Company (*Vereenigde Oost-Indisch Compagnie* or VOC) in the Indonesian archipelago and the Spanish government in the Philippine islands had significant ramifications on the horse populations of maritime Southeast Asia. The various horse breeds of Indonesia were traded from one end of the archipelago to the other, though Java was a favourite market given that island's disproportionate concentration of centres of power and population. Horses were also traded to Siam and southeastern India.[85] Given the global networks of trade and administration of these European empires, however, horses were subsequently diffused on a much larger scale, colonising the Philippines and even extending across the Indian Ocean where they were also introduced into southern Africa.[86] This 'invention' of new horse breeds involved the dissemination of equine genes and phenotypes from Europe, Asia and America and their admixture through purposeful state intervention and inadvertent natural miscegenation. The resultant breeds were as much the product of the changes brought about by their adaptation to new environments as it was the consequences of their close interconnectedness with the wider tides of human historical development. Moreover, far from being shielded from the effects of increasing globalisation and technological innovation, the horse's livelihood came to be increasingly threatened by both externally introduced disease that severely reduced its numbers and by colonial science that heralded an unprecedented curtailment of its reproductive capabilities.

Southern Africa

Unlike research on horses in Southeast Asia which suffers from an absence of records, there is a vast and largely untapped vein of primary source material, generated by anxieties over early settler importation of the equids, the difficulties of using horses in military conflicts, the rise of the horse racing industry and the enduring controversies over the legitimacy of various indigenous breeds.[87] While the horse was in widespread use in North and West Africa from a millennium before, there were no domesticated horses in southern Africa prior to European colonisation as the tsetse fly created an effective barrier to their further penetration. Just as in the case of Southeast Asia, the growth of the colonial state and the rise of the VOC during the seventeenth century were instrumental in the introduction and expansion of equine populations.

As part of an attempt to entrench their mercantile interest in India and the Far East, the VOC established a re-provisioning station at the Cape in 1652. Land was apportioned to a free burgher community with the intention of creating independent commercial farms that would provide the settlement with a steady food supply. In aspiring to reshape the landscape, these first white settlers attempted to transform the physical environment through the importation of horses.[88] Horses were the first domestic stock imported by the settlers after their arrival at the Cape of Good Hope and horses became integral to their identity as Europeans, used both symbolically and in a material sense to affirm white difference from the indigenous population (see Swart this volume). The long journey between the Netherlands and the Cape militated against sending Dutch horses and so the VOC had recourse to sending stock from their base in Java (probably from Sumbawa). These Javanese imports were small and hardy creatures, thirteen and a half hands high.[89] Subsequently known as 'South East Asia Ponies', they were an amalgam of Arab and Mongolian breeds whose ancestors had been acquired from Arab traders in the East Indies.

Such horses contributed significantly to the establishment and growth, particularly in the realms of agriculture, the military and communications, of the early modern colonial state, which began to prosper even against resistance from the metropole. As in the case of Southeast Asia, the 'invention' of new horse breeds meant the dissemination of equine genes and phenotypes from Europe, Asia and the Americas and their fusion through deliberate state intervention, individual efforts by groups of breeders and by simply economic and geographic expediency. The resultant breeds were thus partly a product of their adaptation to new environments, and largely a corollary of their close connection to human society. The breeds became increasingly differentiated, moving from the stock of small 'South East Asia

Ponies' to include English Thoroughbreds imported for the racing industry subsequent to British imperial interest in southern Africa from the late eighteenth century. The English Thoroughbred followed a very different trajectory from the 'South East Asia Pony' stock or 'Cape Horse' with its globalised admixture of American and European stock. Different breeds came to be representative of, and occasionally to represent almost metonymically, different colonial identities and the rise of nationalisms.[90]

This Book

The aims of the chapters that follow are twofold. First, to show how colonialism had repercussions beyond those usually associated with its impact on colonial peoples and environments and also had long-lasting consequences on animal populations. The second aim is to give greater prominence to the role of introduced species like the horse in historiography, how it fared in its new environment and its changing role in colonial societies. In particular, the book explores the 'invention' of specific breeds of horse in the context of imperial design and colonial trade routes. The Cape Horse in South Africa, the Basotho Pony in the mountain kingdom of Lesotho, the Philippine Horse in that part of maritime Southeast Asia and horses in Thailand share a genetic lineage with the horse found in the Indonesian archipelago. The ships of empire carried not just merchandise, soldiers and administrators but also equine genes from as far afield as Europe, Arabia, the Americas, China and Japan. In the process, they introduced horses into parts of the world not native to that animal in historical times.

Divided into two sections, the study deals first with the royal origins and evolution of horse breeds within the Indonesian archipelago, the regional trade in horses between the VOC based on Java and the court of the Siamese (Thai) king, and sketches the outlines of the European shipping routes that carried these animals across the Indian Ocean. These chapters mainly focus on shifts in the commercial and increasingly imperial ties that linked breeders and courts to traders and buyers and disseminated equine genes across the seas. The scenario shifts somewhat in the next section to include both new players and other lands. The Spanish colonial government in the Philippines also became a dealer in equine genes with links not only to Northeast Asia but also as far afield as Europe by stretching back across the Pacific via Mexico. The narrative then takes the horse to southern Africa where its evolution alongside both European and indigenous societies is explored in relation to national identity and state formation. The emphasis here is more on the way horses can be seen as significant agents of historical change in their own right, influencing the indigenous cultures they came in contact with and adapting to the new environments they encountered. To tell this story that spans oceans, continents, centuries and (perhaps widest of all) epistemological and academic boundaries has necessitated more than one author. This is not an edited volume as such as it tells a continuous chronological tale about a single subject, jointly written between two principal authors and four contributing ones. The resultant text is all the richer for its varied expertise and different perspectives.

The first section opens with an account of the Indian Ocean horse trade written by William Gervase Clarence-Smith. Here the scale and extent of the regional horse market is revealed and the periodic changes in supply and demand detailed. He outlines the close relationship between horses and colonial commodity production, mining in South Africa and plantation economies in the Mascarenes (Mauritius and Réunion) and Southeast Asia. By the nineteenth century, however, the horse trade in the latter had become more localised and given way to South Africa as a major regional supplier though epizootics (African Horse Sickness) and warfare (Second Anglo–Boer War 1899–1902) had largely put a stop to exports by 1900. The next two chapters chart the evolution of horse breeds in the Indonesian archipelago. Peter Boomgaard links horse breeding to the political, economic and cultural developments in the areas where they originate, demonstrating the key role that royal courts seem to have played in their gestation, and outlines the trade within and beyond the archipelago. At the same time, though, he emphasises just how 'loose' a term 'breed' really is and how difficult it is always to identify one from the available historical sources. Bernice de Jong Boers then presents a detailed study of one of these famed horse breeds, that of the island of Sumbawa sometimes referred to as the 'Arab' of the archipelago. She shows the relationship between natural environment, mode of human occupation and horse breeding and begins to assess the ecological consequences of this activity. She also makes the telling point that while horse breeding may have been suitable for Sumbawa, the island's environment was not necessarily conducive to the animal's flourishing, so recognising the importance that needs to be paid to non-human historical perspectives. This section concludes with an account of Javanese horses at the court of Ayutthaya in Siam. Dhiravat na Pombejra details one of the earlier trades in horses where the exchange was as much a factor of inter-state relations as it was a market transaction. In the course of this history, a fascinating portrait of the role of horses in seventeenth and early eighteenth century Siamese society is exposed. Again the link between court and breeding is made but now it is also the colonial state in the form of the VOC that emerges as the dominant player.

Still in Southeast Asia, the focus in the initial chapters of the second section shifts northwards to the Philippines. Here the horse was a new-comer, more of an 'invader' than the Spaniards who introduced it into most of the islands through bringing stock from China, Japan and as far afield as Mexico. Greg Bankoff first depicts how the animal was forced to adapt to the new circumstances and environments it found itself in and how its life was increasingly altered by its close association with and integration into colonial society. While sufficient inter-breeding over generations blended the animal's different pedigrees to create a native Philippine Horse as a distinctive breed, Bankoff further argues that the horse did not remain unchanged by these processes but adapted over time through attenuation. The colonial state's subsequent attempt to increase the size of the domestic horse in the Philippines (as in other areas of South and Southeast Asia) through out-breeding local mares with imported stallions from British India and Australia also reveals the way in which Europeans had subtly begun to redefine their own relations to the natural and human worlds about them by the late nineteenth and early twentieth centuries.

The final chapters take the reader on a voyage across the Indian Ocean to the shores of southern Africa. Ponies imported by the Dutch from their Southeast Asian colonies, were critical in the socio-political settlement of the Cape. After their introduction, horses were used as draught animals to effect changes in the new environment, and also utilised by the authorities as a signifier of 'difference', particularly to emphasise the difference between native and settler, in order to assist in the psycho-social subduing of the indigenous population. From 1797, two distinct horse cultures emerged – one followed the imperial British-led racing industry, the other a more utilitarian use of horses. There followed a conflict of horse cultures, between those who followed metropolitan fashions and those adopting 'indigenous' settler modes. Moreover, the divergence led to a clear phenotypical difference between the race horses, which were of English Thoroughbred type, and the utilitarian horses, which came to be considered a definite 'breed' known as the Cape Horse, at first not considered of high status but later invested with the pride of (white) settler society. Horses, however, could not be contained within the confines of white settler society. To illustrate the rise of an indigenous equine culture, Sandra Swart discusses the introduction and transformative effect of the horse in Lesotho. The development of the Basotho Pony was as much due to its anthropogenic-natural selection from the original 'Southeast Asian stock' as it was to colonial scientific breed improvement. In the process, the 'Basotho Pony' came to be seen as a symbol of (male) national identity.

The chapters that now follow sketch an intriguing tale of sex, violence, change and adaptation in the Indian Ocean world between the sixteenth and twentieth centuries. Though this story is inevitably part of a larger historical narrative, it is one, however, in which the horse figures as the principal protagonist even if its agency is often circumscribed by human agendas.

PART ONE

2 Southeast Asia and Southern Africa in the Maritime Horse Trade of the Indian Ocean, c. 1800–1914

William Gervase Clarence-Smith

Introduction: Two Regions

Southeast Asia and Southern Africa both began by exporting equids over long distances across the Indian Ocean, and yet they both came to concentrate on regional markets towards the close of the nineteenth century.[1] At one level, this reflected relatively limited ecological prospects for breeding, in a context of accelerating local economic growth. At another level, Australia emerged as a formidable competitor, dominating the long-distance market for sea-borne horses of high quality from the Cape of Good Hope to Siberia.[2]

The disease environment stimulated regional trade at both ends of the Indian Ocean. Hotter parts of Southern Africa were infested with tsetse flies, transmitting trypanosomes, blood parasites that caused the deadly disease known as Nagana. Even areas free of this plague were episodically devastated by African Horse Sickness, a virus borne by tiny gnats. Parts of Southeast Asia were similarly infested with Surra, trypanosomes spread by biting flies, and only slightly less dangerous to equids than Nagana.

Regional specialisation meant that Southeast Asia and Southern Africa rarely competed much in overseas markets. Sumatra and the Philippines initially sent modest supplies of ponies by sea to India, the ancient magnet for maritime traffic, but this flow tended to dry up as imports from the Cape Colony expanded in the early nineteenth century. In the Mascarene islands (Mauritius and Réunion), the second largest Indian Ocean market, imports of South African horses faded away from mid-century, just as those from eastern Indonesia began to pick up.

As regional markets came to predominate, transport by sea also became less common. Horses bred in the cooler parts of South Africa were increasingly driven overland, to cool plateaus elsewhere in the region. Similarly, in Mainland Southeast Asia, overland movements predominated, from northern mountains to lower-lying zones. However, geography dictated that large numbers of horses continued to move by sea around Maritime Southeast Asia, notably from Indonesia's southeastern islands to Java, from Sumatra to Malaya, and from many parts of the Philippines to Manila.

South Africa exported some mules and donkeys, whereas these close cousins of the horse were rare in Southeast Asia. The greater incidence of disease in Southern Africa may in part account for a more substantial number of mules and donkeys, but scarcely seems able to explain it altogether. There was probably cultural resistance to owning donkeys and mules in Southeast Asia, inherited from the region's Hindu past, which may explain why mules were largely confined to areas bordering on China. Donkeys were thoroughly impure animals for Hindus, owned by Dalits (untouchables) who transported defiling substances, whereas horses were counted among the most noble of beasts. Mating the two was seen as contrary to nature, especially as the result was a mule or a hinny, neither of which could produce fertile offspring.[3]

A Strategic Business

Equids were highly significant in the economies and societies of the nineteenth century Indian Ocean and South China Sea. Horses were central to many forms of warfare, whether for cavalry, mounted infantry, field artillery, or the baggage train. They were also employed intensively for urban transport, prior to the spread of the internal combustion engine. Many expanding leisure pursuits depended on horses, especially riding, hunting, and racing. Together with mules and donkeys, they plodded along some of the rural feeder routes to railways and river ports, and contributed to farming and forestry. However, bovids (oxen, buffaloes and yaks), or even elephants and camels, were often preferred for rural transport and productive activities. Horses were occasionally eaten, depending on local food taboos, but were only specifically bred for meat and milk in Mongolian and Turkic parts of Inner Asia.

Commercial intelligence and plentiful capital were crucial to success in the maritime trade in equids. Prices fluctuated rapidly in response to demand, and supply factors were difficult to predict. Because of its strategic nature, the trade was monitored by governments, which sometimes intervened somewhat clumsily, as either purchasers or regulators. Shipping required considerable amounts of working capital, for live animals needed much care and attention on board ship. Returns could be very substantial, but risks were high, especially prior to the introduction of steamers. Food and water might run short, notably if ships were becalmed, epizootics could easily break out, and poorly treated or storm-tossed animals could run amok.

Around 1910, a very roughly estimated 30 million equids, or a little under a fifth of the world's total, were scattered around the lands bordering on the Indian and Western Pacific oceans, excluding Inner Asia. Southeast Asia accounted for about 1,111,000 horses, compared to around 842,000 horses and 480,000 mules and donkeys in Southern Africa.[4] Equids were mainly bred in dry and temperate areas, conducive to fertility and health. Level, lightly populated, and calcareous lands allowed for good muscle and bone formation. Cool uplands and dry zones were often favoured in these tropical and subtropical climates.

India and China had for centuries been the 'sinks' for horses in the region, drawing their supplies mainly by land from the teeming herds of Inner Asia. A maritime trade had gradually come to supplement overland flows, with the Persian Gulf as the main supplier and India as the main importer. To a lesser extent, the lower reaches of the Red Sea, Burma, parts of Maritime Southeast Asia, the Ryukyu Islands and Korea also exported horses by sea.[5]

It was not until the late eighteenth century that South Africa began to enter these maritime markets. Soon after they settled at the Cape in 1652, the Dutch brought horses from Persia and Java, and fresh breeding stock later came from Europe and the Americas.[6] Donkeys from the Cape Verde Islands arrived in the eighteenth century, giving rise to some mule breeding.[7] Horses passed into indigenous hands from around 1822, and the Sotho became the most successful indigenous Africans in this domain.[8]

Australia apparently obtained its first horses from South Africa in 1788, and they throve in a highly favourable environment, containing few diseases or natural predators. There were successive additions of English, Arab, Timorese, and Chilean blood. The name Walers, derived from New South Wales, was progressively applied to horses of every breed, type and origin, sometimes even including those from New Zealand. Exports only really took off in the mid-nineteenth century.[9]

In terms of imports, the nineteenth century exhibited a mix of continuity and change. India consolidated its position as the main maritime destination for equids, coming mainly from the Persian Gulf, South Africa and Australia. The Mascarenes, developing intensive sugar cane production, were relative newcomers, seeking equids from an astonishing variety of regions, namely Australia, Eastern Indonesia, the Persian Gulf, Ethiopia, South Africa, the River Plate, North Africa, and Europe. Continental Southern Africa gradually shifted from net exporter to net importer, as a result of the 'mineral revolution' of the 1860s. Southeast Asia's towns increased their horse consumption greatly, usually tapping into regional resources. East Asia played little part in the maritime trade in equids, as China was served essentially by overland routes. However, Japan occasionally entered the market quite aggressively, notably in times of war.[10]

The Rise and Fall of South Africa's Maritime Exports

South Africa initially benefited from a major British strategic headache, the drying up of India's traditional Inner Asian sources of equids. This was only partially compensated by increasing imports from the Persian Gulf, especially as most Arab horses were fairly small.[11] To make matters worse, there was a marked fall in horse and mule breeding in northwestern India, probably due to the increasing cultivation of pasture land and the disruption of trans-humance routes.[12]

The first recorded shipment of 'Capers' to India was in 1769, but exports remained spotty for decades.[13] After the British had secured possession of the Cape in 1814–1815, Lord Charles Somerset contacted Madras Presidency officers about increasing exports. In 1816, he proposed supplying India with several hundred horses a year, transported on East India Company ships. The Company demurred, and exports continued to be sporadic, often limited to animals bought at the Cape by British personnel bound for India.[14] Nevertheless, purchases were stimulated by the success of South African horses on the Calcutta race track from 1812, and their popularity with British officers in India.[15] Thus, the horse trade slowly became significant for certain South African farmers.[16]

South African exports to India became more regular from the 1840s, although rather high prices limited the flow to some 50 a year from 1841 to 1850.[17] At the time, Cape ports were sending about 200 horses a year to Mauritius.[18] Capers were only a partial solution to British problems in India, however. To be sure, they were tough and well adapted to tropical conditions, making excellent cavalry remounts. They were somewhat undersized for horsed artillery, however, even if taller than Persian Gulf and 'native' breeds. A team of eight 15 hand Capers was allocated to a six-pound gun, whereas six horses would have been usual in Europe.[19] Capers were chiefly bought in Bombay and Madras presidencies, but also reached Calcutta and Ceylon.[20]

The 'Indian Mutiny' of 1857–1858 brought exports to a brief peak. British military agents bought 5,482 horses and 108 mules in South Africa from 1857 to 1861, generating a windfall gain of £215,645 in export revenues.[21] The British also procured a similar number of horses from Australia. Symbolically, Sir Henry Havelock made his official entry into Lucknow on a South African mount, while the legendary George Hodson rode an Australian horse at the relief of Delhi.[22] From then on, British officials sought to privilege regular supplies from these imperial territories, carried on steamers.

Despite this, South Africa's overall horse exports fell away markedly from the 1860s, as suggested in Table 2.1. Even when they seemed to grow again in the early twentieth century, they were more or less balanced by imports, and net imports of mules stood at around 2,500 a year in 1911-1914.[23] Some factors put forward to explain this trend are far from convincing. Warfare in Southern Africa only briefly constrained supplies. Moreover, a healthy trade should have spawned its own transport, even if less shipping was temporarily available after the opening of the Suez Canal in 1869.[24]

Disease might appear to have been a more serious problem, although there are difficulties with this explanation as well. African Horse Sickness occasionally surged southwards, ravaging South Africa with epizootics approximately every 20 years. A severe outbreak in the mid-1850s was estimated to have killed nearly 65,000 horses.[25] Such epizootics had a transitory impact, however, for the horse population of the Cape and Natal rose from 145,000 in 1855 to 446,000 in 1899.[26] This increase also casts doubt on alleged competition for pasture from merino sheep, goats and ostriches.[27]

South Africa's 'mineral revolution' constitutes a much more convincing explanation for falling exports from the late 1860s, as internal prices rose to the point that Capers ceased to be competitive in India.[28] Indeed, breeders no longer strove to meet the requirements of Indian artillery units, since they could easily find local purchasers for smaller horses.[29] Although plans to send Caper remounts to India lingered on till the 1890s, the trade was over.[30] The use of steamers from the early 1880s cemented Australia's advantage on the Indian market, eliminating many of the delays and dangers that had attended the trade in sailing vessels over a longer distance and with less predictable winds.[31]

Table 2.1 Horse Exports from Southern Africa to all Destinations, 1827-1918.[32]

	Cape ports only	South Africa
1827	319	
1833–39 average	284	
1841–50 average	234	
1857–61 average		1,096
1865	98	
1870	175	
1875	29	
1880–85 average	90	
1911–18 average		788

Demand within Southern Africa was varied. 'Horse-whims', large wooden wheels powered by equids, raised ore from open diamond pits in Kimberley.[33] More significant was surging demand from transport riders. For people and goods needing to go at a faster pace than ox-wagons, entrepreneurs harnessed horses, donkeys, mules, and even domesticated zebras.[34] The progressive opening up of the continent also redirected South African exports overland, notably to the Rhodesias, South West Africa, Angola, and Mozambique.[35]

The evolution of the multinational Vanrenen family reflected these changing trends. Originally Prussian aristocrats exiled to the Cape, they fitted their name into a Dutch spelling, and became anglicised. One family member was involved in breeding horses for the Indian market in South Africa in the early nineteenth century. Henry Vanrenen began exporting Australian horses to India from the middle of the century, helped by his brothers serving as cavalry officers. It was this latter branch of the business that became crucial to family fortunes, whereas South African operations declined.[36]

The Second Anglo–Boer War (1899–1902) brought an unaccustomed flood of imports. The British deployed some 494,000 horses against the Boers, of which about 334,000 were imported, nearly half from the United States. In addition, some 67,000 mules arrived from the United States. Altogether, about a tenth of the total cost of the war was attributed to the purchase of equids, in part because of astounding losses of some 326,000 horses, very few of which were killed in action.[37] Other military campaigns led to smaller bursts of imports into areas free of tsetse, notably Madagascar, the Horn, and the Nile valley.[38]

Conversely, South African equids bore the brunt of African campaigns in the First World War. The recently created Defence Force had some 8,000 horses and mules at the outbreak of hostilities, a figure which rose to around 160,000 by January 1916. After suppressing a Boer rebellion, the South African army seized German South West Africa in 1915. However, attempts to repeat this success in German East Africa in 1916 turned into an equine holocaust, due to the ravages of Nagana. Out of 31,000 horses, 33,000 mules and 24,000 donkeys sent there, only 3,108 animals remained alive in October 1916.[39]

Plantation Markets in the Mascarenes

The islands of the Southwestern Indian Ocean were the largest purchasers of equids in relation to their human population, benefiting exporters in both South Africa and Indonesia. The cultivation of sugar cane required many animals for cartage; there were horse-powered sugar mills in early years, and urban transport grew.[40] Indeed, the British urged the use of horses to cope with problems caused by the restriction of slave imports into Mauritius from the early nineteenth century.[41] Not until the spread of light railways around 1900 did the demand for equids begin to falter.[42] Even then, a passion for horse-racing maintained demand.[43]

Mauritius and Réunion raised hardly any equids of their own, even though the Dutch had brought horses to Mauritius in 1666.[44] Donkeys came in the eighteenth century under the French, who also took horses on to the Seychelles.[45] The problem was that there was little land or labour available for breeding, so that periodic attempts to cut import bills met with slight success. The islands were free from Nagana, but equids suffered from Surra.[46] Disease might explain why mules and donkeys were more numerous than horses, as they were probably slightly more resistant. However, this could also have reflected the French background of most settlers, accustomed to these animals.

Whereas cattle could easily be obtained from neighbouring Madagascar, equids came from many distant lands, beginning with the Persian Gulf. Oman was said to be already sending mules to the French plantations in the Mascarenes in the eighteenth century.[47] In the early nineteenth century, Oman specialised in exporting large donkeys to Mauritius, though these were mainly raised in Bahrain and its hinterland. The trade picked up in the late 1830s and averaged some 630 a year in the 1840s, with a few horses to complement it. Although this business was reported to be 'extinct' by the 1870s, minor imports of 'Muscat asses' persisted to the late 1890s.[48] These may in reality have been drawn from East African coastal areas, for donkeys were raised at scattered points from Somalia to Mozambique, and were sometimes exported.[49]

Table 2.2 Mauritian Imports of Equids 1833–1913[50] (Annual Averages; Some Years Missing; 1851–1860 Decade Missing).

	Horses	Mules	Donkeys	Equid total
1833–39	423	1,040	(with mules)	1,463
1841–50	403	1,553	(with mules)	1,956
1861–70	416	1,092	30	1,538
1871–80	650	2,079	79	2,808
1881–90	499	966	142	1,607
1891–1900	371	948	61	1,380
1901–10	741	477	82	1,300
1911–13	286	102	0	388

In the 1830s and 1840s, British trade reports showed some 500 mules a year entering Mauritius from France, a major exporter in world terms, with some competition from the Horn of Africa and South Africa. A ship or two from Mauritius and Réunion called at the Eritrean port of Massawa each year, bringing around 400 a year to the Mascarenes. Mules increased in value by a factor of more than ten between Massawa and the islands, but risks and costs were considerable. The animals were out of condition on the hot coast, and thus highly prone to disease. Ships had to provide ventilation, water, fodder, and protection against injury from rolling and pitching. The trade withered away from the 1870s, possibly because local regional demand was rising, notably in Ethiopia, but also in the Sudan and Kenya.[51]

South America then became the main supplier of mules to Mauritius, although the origins and dynamics of this trade remain to be uncovered. As early as 1834, 26 mules arrived from 'The States of the Rio de la Plata', and the trade may well have begun earlier.[52] Argentina replaced Uruguay as the chief source of mules in the late 1890s.[53] Complicating the picture was a late nineteenth century re-export trade in mules from Mauritius to India, of uncertain size.[54] France and Algeria probably kept a significant position on the Réunion market, as French protectionism increased from the 1880s.

South Africa and Australia provided horses for Mauritius in roughly the same sequence as for India, with South Africa entering the market earlier, but dropping out after the mineral revolution.[55] Western Australia specialised in this niche market for a time, probably exporting timber with horses, but the 1890s gold rushes put an end to the business.[56] Australian Walers continued to dominate the Mauritian turf, however, until just before 1914, when the island turned to European racehorses.[57]

The Lesser Sunda Islands of Indonesia found a fairly steady outlet for their cheap little ponies in the Mascarenes, probably growing out of earlier exports of slaves. The first certain mention of this trade dates from 1821, and it remained firmly in the hands of French ships' captains, with Mauritius a larger customer than Réunion.[58] Just over 100 horses a year arrived in Mauritius from the Dutch East Indies in the 1840s, and the Lesser Sundas provided between a third and two thirds of the island's imports from the end of the 1870s to 1914.[59] Sailing ships from Mauritius visited the Lesser Sundas till 1914, but did not return thereafter.[60] Réunion seems to have dropped out earlier, probably after the French imposed protectionist duties in the 1880s.[61]

Internal Markets in Southeast Asia

Southeast Asia's horse trade came to be overwhelmingly focused on the region itself.[62] In the seventeenth and eighteenth centuries, and possibly before, Muslim Tamil traders controlled a modest flow of North Sumatran ponies to Southeastern India, a trade that was linked to a parallel one in elephants.[63] Horses from the Philippines were also reaching the same part of India from 1708, if not earlier.[64] This business withered away in the early nineteenth century, possibly because of increased demand in the Philippines and Malaya.

Maritime exports from Rangoon, mainly directed to Northeastern India, lasted longer. About a hundred horses a year were exported in the 1870s, with some beasts reaching Southeastern India and Malaya. They were sold under the name of 'Pegu Ponies'.[65] This name was misleading, for Pegu was no centre of horse breeding. It has been suggested that these horses were really raised in the Shan States of northern Burma.[66] However, most of them seem to have originated from Yunnan and eastern Tibet. They were brought to the Irrawaddy at Bhamo, and shipped down river to Rangoon, whence some of them were sent further afield.[67] The active horse trade of Mainland Southeast Asia was otherwise conducted overland.[68]

Exports to East Asia scarcely fared any better, in part because China bred more equids than India, and was more successful in procuring supplies from Inner Asia.[69] Southeastern China, poorly situated for both overland supplies and local breeding, was the most promising market. Horses entered this region from the Philippines and Timor, reflecting ancient Spanish and Portuguese connections with Macau, but both trades declined over time.[70] Australian Walers supplanted Southeast Asian ponies from the 1840s, and also penetrated Japan to a limited extent after the Meiji Restoration of 1868.[71]

Malaya, undergoing a tremendous plantation boom from the 1880s, bred no equids, although oxen and elephants were prominent in rural transport.[72] Annual imports of horses thus ran at around 2,000 in 1895–1897, probably growing to some 3,000 a year by 1905. Singapore and Penang acted as entre-pots, redistributing animals to smaller ports.[73] Imports came mainly from Indonesia in Dutch-flagged ships, owing to Dutch protectionism, although the shippers were rarely Dutch.[74] Batak, Gayo and Minang ponies from the Sumatran highlands were misleadingly known as 'Deli ponies' in Malaya, as they were shipped from East Coast ports in the Deli area.[75]

Sumatra remained the principal source for Penang, a trade amounting to about 500 head a year in the late 1890s and early 1900s, whereas Singapore imported more 'Java ponies', possibly re-exported from the Lesser Sunda Islands.[76] 'Java ponies' provided most of the draught power for Singapore's numerous cabs up to the 1880s.[77] They were also the staple of early horse-racing, after early experiments with Burmese and Chinese horses. However, Singapore turned increasingly to Australian supplies from the 1890s.[78]

Java's towns attracted numerous ponies from the Lesser Sunda Islands, even though there were local breeds, including the beautiful little Kedu horses of central Java.[79] Indeed, the Priangan highlanders bred the largest horses in Southeast Asia, at around 15 hands, sending some 3,000 a year to the Buitenzorg (Bogor) market in the 1920s.[80] Among the Lesser Sundas, Sumbawa already exported more than a thousand horses a year prior to the catastrophic eruption of Mount Tambora in 1815.[81] South Sulawesi's relatively large beasts were prized by the Dutch armed forces on Java, but disease took its toll, and exports declined from the 1860s.[82] The golden age for Sumbawa and Sumba was from the 1890s to the 1910s, whereas exports from Bali, Lombok, and Timor fell away.[83]

Official protection of Dutch-flagged shipping made it virtually impossible for the British to break into the carrying of horses from the Lesser Sundas to Java, which long remained an Arab preserve. From 1818, all coastwise shipping was legally restricted to Dutch vessels, owned by Dutch subjects. British shippers lost market share, initially to the benefit of Hadhrami Arabs, legally Dutch subjects, who owned many European rigged ships.[84] For several decades, Arab ships fought off competition from Dutch steamers, introduced in the 1860s.[85] However, Arab shippers lost out when special facilities for feeding and stabling live animals were introduced on steamers around 1900.[86] The Koninklijke Paketvaart Maatschappij, founded in 1888 for inter-island services by a cartel of Dutch shippers, then became the chief transporter of Lesser Sunda horses to Java, benefiting from generous official subsidies and assistance.[87]

Arab domination of the trade outlived their loss of control over shipping.[88] They already enjoyed a commanding position by the 1840s in Sumbawa and

Table 2.3 Horse Exports from the Lesser Sunda Islands, 1880–1918[89] (Annual Averages by Decades; Very Approximate; Mostly to Java).

	Bali & Lombok	Sumbawa	Sumba	W. Timor & depend.	Total
1821–30	n.a.	c.200	–	129	
1831–40	n.a.	c.250	–	248	
1841–50	n.a.	c. 1,000	680	336	
1851–60	c.1,000	c. 4,000	n.a.	597	
1861–70	n.a.	c. 2,000	c. 750	1,187	
1871–80	n.a.	n.a.	c. 2,000	2,966	
1881–90	c.350	n.a.	1,851	1,120	
1891–1900	c.100	c. 6,000	c. 3,000	c. 750	c. 10,000
1901–10	c.10	4,357	1,302	c. 750	c. 6,500
1911–20	c.1,000	5,065	2,520	c. 2,000	c. 10,000

Sumba.[90] The trade was 'completely in their hands' in Java by the 1880s.[91] Arab merchants were typically transients at this stage, sending junior members of family firms based in Java to secure cargoes.[92] In the early 1870s, Arabs on Sumba built temporary houses made of bamboo or mats, with simple enclosures for horses. After a lengthy physical examination, bargaining went on for hours.[93] As Arabs increasingly took up residence in the Lesser Sundas, they tightened their grip. Some went into the interior to purchase horses, but they usually worked through local brokers. Advances in trade goods, purchased on credit in Java, were the norm, and the final settlement of accounts followed sales in Java. Horses were either delivered directly to Dutch military depots to fulfil contracts, or were sold by auction.[94]

Javanese imports of foreign horses were on a smaller scale, but they grew over time, typically involving Australian Walers for the army and wealthy Europeans.[95] Arab horses from the Persian Gulf were repeatedly investigated by Dutch cavalry officers. However, they were never considered suitable, mainly because their average size of 14 hands was too small.[96] Breeders in Australia's Northern Territory were geographically well placed to break into the Java trade by exporting from Darwin from the 1880s, but they met with little success.[97] Western Australia did better, with exports to Indonesia and Malaya peaking at 3,109 in the decade 1871-1880, but falling dramatically in the 1890s, as all horses were required to service the gold rushes.[98]

Eastern Australia's fortunes improved when the Dutch cavalry decided to replace Lesser Sundas ponies in 1902. By 1907, the whole force was mounted on Walers. However, this was a much smaller market than that of British India, for Dutch cavalry units were largely confined to Java, and only disposed of around 700 mounts at this time. Moreover, after entering the war in 1914, Australia banned exports of horses to the neutral Dutch possessions, leading to hasty purchases of unsatisfactory animals from China. The ban was temporarily lifted in 1916, but only for a batch of 250 horses, all of which were under 14 and a half hands.[99]

Manila was Southeast Asia's second great mart for sea-borne horses.[100] The city relied on animals from all over the Philippines, with Chinese and Chinese Mestizos as the main horse traders in the 1840s.[101] From areas at a certain distance, animals came by sea, usually on sailing craft. In 1862, 177 horses arrived this way, just over half from Zambales and Ilocos to the north. The rest nearly all came from the islands of Mindoro and Lubang to the southeast, or the neighbouring Bikol peninsula.[102] The Spaniards reserved coastwise shipping to their own flag, although British and other Western firms employed men of straw.[103] The Philippines and Guam obtained few horses from foreign sources before the end of Spanish rule in 1898, although some Australian Walers were sent to Manila in the 1880s.[104]

After Spain's defeat in 1898, the Americans faced a crisis in the supply of horses. The archipelago's livestock resources were ravaged by guerrilla warfare. Diseases, notably Surra, were either introduced by American troops, or were spread more widely by military campaigns to subdue arduous Filipino resistance. Prices roughly tripled in a decade. The colonial government imported thousands of horses and mules from the United States, but North American equids were costly, spurned local fodder, and had no resistance to Surra. Private American dealers thus turned to cheaper and hardier Walers.[105] American officers eventually followed suit, with military purchasing agents stationed in Australia in 1904. By 1910 about a thousand Queensland horses had been sent to the Philippines, largely for military use.[106]

Conclusion

The First World War marked the swan-song of the global trade in equids. The conflict moved animals and fodder around the globe on a vast scale, and losses were colossal, in part because of heroic but vain cavalry charges against barbed wire and machine guns. This equine catastrophe acted as a tremendous spur to the development of mechanical forms of transport, hastening the triumph of the internal combustion engine.

The displacement of horses from their two main niches in Africa and Asia was slower than in the West, as cars, lorries and tanks took longer to become established on the periphery. Equids thus continued to be shipped in the inter-war years, albeit in declining numbers. The overland trade, notably from Yunnan and Tibet to Mainland Southeast Asia, proved more resilient. The sea-borne trade's most enduring aspect was the commerce in race-horses and polo ponies, but today these luxury animals are frequently conveyed by air.

The twentieth-century eclipse of equids by automotive vehicles has obscured the importance of these beasts in the previous century. In turn, this has led to a neglect of their role as a minor but valuable and strategic component in the sea-borne trade of the imperial era. The Indian Ocean and the China Sea contained a loosely interlinked and overlapping set of regional maritime markets, of which Southeast Asia and Southern Africa were only two. More research in needed to understand the internal dynamics of these regional trades, and how they fitted in with one another.

3 Horse Breeding, Long-distance Horse Trading and Royal Courts in Indonesian History, 1500–1900

Peter Boomgaard

Introduction

In the historical sources and the literature on the Indonesian Archipelago dating from the nineteenth and early twentieth centuries, a rather limited number of horse breeds are usually mentioned. There is some variation between sources, probably largely due to changes over time, but when that is taken into account, the similarities are more striking than the differences. This chapter attempts to relate the existence of these breeds to political, economic, and cultural developments in their areas of origin.

There is, however, one principal difficulty and that is that none of the authors considered here has defined precisely what he means by a 'breed' (often called 'race'). In principle a breed is a subspecies that can be distinguished from other subspecies by its external characteristics, such as height, colouring, and build. But authors rarely specify the criteria they have used when they, for instance, distinguish two breeds in a certain area, while someone else records only one and yet another observer reports the existence of two varieties within one breed, or even two 'sub-breeds'. This is by no means purely a problem of the past; modern biologists recognise that there are 'splitters' and 'lumpers' within their ranks. Such decisions are apparently based on subjective inclination rather than on rigid adherence to 'scientific' rules, as discussed in this volume's introduction. For the type of analysis presented here, this is not an insurmountable problem, as will be indicated below.

First a few general points about horses in pre-modern Asia. The most important point might be that all Southeast Asian horses are small. Large horses, therefore, were at a premium, and they were perfect gifts to be bestowed upon or presented by rulers. Consequently, we witness a constant stream of horses from and to Batavia (present-day Jakarta), headquarters of the VOC (Dutch East India Company), which thus became the hub of the inter-Asian exchange of horse genes. Presenting good horses to rulers was so important that the VOC, apart from buying and exporting fair numbers of very expensive (large) Persian and Arabian horses from Mokka and Gamron (from the 1640s), also started its own 'stud farm'. This was done on a number of small islands off Jafnapatnam [Jaffna], on Ceylon [Sri Lanka]. Apparently the Portuguese had been using these islands for a similar purpose, as they called them Ilhas de Cavalhos [Horse Islands] and one of them Ilha das Vacas [Cattle Island].[1] From 1666 onward, horses from Java and Makassar [Sulawesi] were sent here, and at least after 1692 also Persian stallions. In 1717 we encounter horses from Bima [Sumbawa] being sent there.[2] In the first decades the stud farm operated at a loss and it was abolished in 1701. In 1705, however, operations revived, and thereafter the stud farm returned a tidy profit, for unclear reasons. The farm even survived losses precipitated by natural hazards, such as the death of 105 horses (out of a total number of 453) in a storm in November 1726. In 1758 there were 647 horses recorded, and in 1762, the last available reference, the stud farm was still doing well.[3]

Thus horses from various geographic origins in Asia were presented to rulers all over Asia. The latter included the Shogun of Japan and the king of Kandy [Sri Lanka], both of whom were very particular about horses, to the extent of giving the VOC representatives in Deshima very detailed instructions on the kind of horses they liked and disliked.[4] It reminds us of the enormously detailed Javanese literature on horses with lucky and unlucky characteristics (*katuranggan*). Such 'horse lore' was a widespread and socially entrenched Asian feature.

A few words are also in order about the VOC, as they are one of the major agents in the story to be told here. During two centuries – from c. 1600 to c. 1800 – the Company made over 8,000 trips with their large Indiamen to and from Asia. In Asia itself, their ships went annually to regions such as Arabia and Persia, and to various harbours in what are now Sri Lanka, India and Bangladesh, where they had many establishments. Other establishments that were visited annually were located in the Malay Peninsula, Sumatra, Java, Borneo, Sulawesi, the Moluccas and the Lesser Sunda Islands, most of them now forming part of the Republic of Indonesia. Burma [Myanmar], Thailand and Vietnam were also visited regularly, as were Taiwan (in the seventeenth century), China (in the eighteenth century) and Japan. 'In the seventeenth and eighteenth centuries the VOC was the largest single employer of European and Asian men in the Indonesian archipelago.'[5] Many people will have heard of the VOC, associating it with the trade in precious spices, such as cinnamon, nutmeg and cloves. This was certainly an important part of their business, but the Company was also involved in the trade in much bulkier items, such as timber, cattle, and even elephants. In this chapter, they will make their appearance as transporters, buyers and sellers of horses.

Sumatra

Horses were introduced in the Archipelago before the ninth century – possibly even as early as the third century [6] – but are not often mentioned in the sources dating from 1500 and earlier. We find nothing in the Chinese sources dating from the fifteenth century, translated and published by Groeneveldt.[7] They mention a range of other domesticated animals in various Sumatran 'kingdoms', but not horses. However, an early twentieth-century author emphasised how much the Batak breed from northern Sumatra resembled the Mongolian horse morphologically, which might point to a Chinese introduction at some point in time.[8] The Portuguese author Tomé Pires, whose writings reflect the situation around 1515, only once refers to horses in Sumatra. This occurs when he deals with Pariaman, a port-of-trade in central western Sumatra. According to Pires, the place had many horses, 'which they go and sell continuously in the kingdom of Sunda'.[9]

Another Chinese source, dating from the early seventeenth century, states that the region around Aceh in northern Sumatra had good horses.[10] It is the only Chinese source in which horses are mentioned in the writings published by Groeneveldt dating from that period. Around the same time (c. 1620), the French merchant Augustin de Beaulieu saw '*beaucoup de chevaux, mais de petite taille*' [many horses, but small ones] in the pastures around the city of Aceh. His testimony is important, because most seventeenth-century travellers only mentioned the many tame elephants and buffaloes to be found there. However, the horses seen by Beaulieu c. 1620, may have all but disappeared by the time the English merchants Peter Mundy (1637) and William Dampier (1688) visited the area.[11] After all, the plain surrounding the city was not all that large, and it is likely that pasture for horses had been turned into ricefields in the mean time.

A Dutch source dating from 1644 mentions ships from India and Pegu [Burma], exporting horses – and elephants – from Aceh. The same source reported that horses were available in Deli, a coastal region to the southeast of Aceh. A Dutch compilation dating from c. 1700 also mentions Indian merchants, but this time they are bringing Persian horses to Aceh.[12]

So was Aceh importing or exporting horses, and was it breeding them in the area adjacent to the city? The Acehnese may have been raising horses at the beginning of the seventeenth century, but later on, when population growth must have made finding space for pasture a problem, they probably no longer did, at least not on a large scale. It seems likely that the rearing of horses had been taken over by the nearby Batak highlands (see below). From there they were probably taken to Deli and then shipped to Aceh, or perhaps there was also a direct overland route. Aceh was probably both importing and exporting. They exported the small but sturdy Batak ponies and imported the bigger breeds, such as Persian horses, no doubt for the ruler and the nobility.[13]

In the nineteenth century, Batak ponies were renowned, but the first printed source to mention them appears to be William Marsden, who published the first edition of his *History of Sumatra* in 1783. He talked about the 'extensive, open and naked plains' which were 'entirely clear of wood, and either ploughed and sown with padi or jagung (maize), or used as pasture for their numerous herds of buffaloes, kine [cows], and horses'. The Bataks seem to have slaughtered horses, specifically reared for that purpose, on public occasions (and to have eaten them). They also hunted deer on horseback and were 'attached to the diversion of horse-racing'. John Anderson, who visited several ports of northern Sumatra between 1823 and 1825, mentioned horses from Aceh, from the Batak country, and from nearby Batubara, Deli and Asahan. Horses were exported from these places, and were 'brought down from the interior in considerable numbers'. It seems likely that the Batak area, particularly the Toba Batak and Karo Batak lands, was the place where many of these horses originally came from. It is possible that after 1900, as the Batak uplands became more densely populated, horse breeding was partly taken over by the adjacent Gayo lands.[14]

So it would appear that the famous Batak horses might have been in existence around 1600, but not yet around 1500. It seems likely that the demand for horses originated from the northern Sumatran courts, of which Aceh was the latest and the one most given to conspicuous consumption. The link between a royal court and the breeding of horses is a recurring theme in this chapter.

In the Archipelago, horses were hardly ever used to draw the plough and, at least prior to the 1800s, were seldom used as beasts of burden. In other words they were not used directly by the peasant cultivators, although smallholders were no doubt called upon to take care of the horses of the nobility, either as a commercial venture, or in statute labour (compulsory services).[15]

Prior to 1800, horses were mainly employed in war, for hunting exploits, in jousting tournaments, and as riding animals for the nobility. Jousting tournaments were held by Central Javanese rulers at least since the early fifteenth century, although originally not on horseback, which suggests that at that point in time horse breeding for or by Javanese rulers may have been of limited importance. From the early seventeenth century onward jousting on horseback was a regular occurrence at Central Javanese royal courts. These lance tournaments would continue until the beginning of the nineteenth century; it seems that at least at the princely courts they did not survive the Java War (1825-1830).[16]

It is tempting to assume that when the coastal plain around Aceh became more densely settled, horse breeding was crowded out, the Batak upland plains taking over its function and becoming the favourite horse breeding pastures. With the Acehnese court having primed the pump, the Batak area, perhaps also better suited to this role than the plains around Aceh, was apparently able to satisfy a growing local and supralocal demand for horses.[17]

But what about Pariaman, recorded as an exporter of horses around 1500 by Pires? According to Christine Dobbin (1983: 71) the place was exporting Batak horses, but if one looks at the map at the end of this chapter (Map 3.1) this does not seem likely. Any outlet for Batak horses would have been located far to the north of Pariaman. The answer is provided by Dobbin's own book.

Horses were to be found in the coastal plain around Padang and in the Minangkabau upland valleys. Nineteenth-century sources mention an Agam breed and a Payakumbuh horse, named after two localities in the Minangkabau uplands, but by analogy with the Batak horse it should simply be referred to as the Minangkabau horse.[18] Dobbin explains that these horses were needed for the transport of gold from the highlands to the coast, a trade that was in the hands of the Minangkabau royal dynasty (or dynasties). One supposes that the courts also employed the horses from the region for royal pomp and splendour, though probably not on the same scale as did the sultans of Aceh. As there had been royal courts in the Minangkabau area at least since the fourteenth century, it is not strange that Pariaman was exporting Minangkabau horses by 1500. The story seems to be very similar to that of the Aceh–Batak–Deli region, apart from the fact that the breeding areas and the courts of the Minangkabau did not become separated.[19]

In c. 1800, the two horse breeding areas, the Batak and the Minangkabau upland valleys, were two of the most densely settled regions of Sumatra. This was no doubt also true in the preceding centuries. The question is, of course, whether there is a causal link between population density and horse breeding, and if so, in what direction the causal arrow is pointing. Horses need vast pastures (think of the Central Asian open plains and semideserts), and in tropical rainforest areas, where vast pastures are not part of the original landscape, fairly large numbers of people are needed to clear the forests and probably also to maintain the open spaces (usually with fire). However, when the people themselves were breeding too successfully and multiplying too quickly, horse breeding may have had to go, as pastures had to give way to arable lands. So it would appear that we have found a negative feedback loop, in which the conditions of successful horse breeding in tropical lowland areas already include the seeds of its demise.

Java

The Minangkabau horses were being exported to western Java, where another court put them to good use. This was the court of Pajajaran, which Tomé Pires called Sunda, a term usually indicating West Java, the area over which the rulers of Pajajaran held sway at that time (c. 1500). According to Pires, 'The land of Sunda has as much as 4,000 horses, which come here from Priaman and other islands to be sold.' There were also 40 elephants 'for the king's array'. The king was a great sportsman and hunter, and he and his nobility went hunting all the time.

In c. 1600, when the sultanate of Banten had taken the place of Pajajaran, the area no longer seems to be a place rich in horses. The king's horse stable was mentioned in passing in one travelogue, but another one dating from around the same time explicitly states that the people of Banten had few horses, and 'small nags' at that. The same source specifies that they are not used as draught animals but purely for riding, presumably by the nobility only.[20]

It would appear that the exports from Pariaman had stopped and that Banten did not produce sufficient horses even to fill its own demand. The Sultan frequently petitioned the VOC in Batavia for horses. Occasionally, high-ranking envoys were sent from Banten to Batavia in order to buy horses as well. The Governor-General of the VOC did send horses to Banten from time to time, including expensive Persian stock. However, one request, made by the Sultan in 1721, could not be met, namely for a 'miracle horse', supposedly roaming Batavia's hinterland. This was the *kuda Sembrani* or Sembrani horse, a fabled animal, supposedly dating from the period of Pajajaran. In the 1840s, the indigenous population of Cirebon was said to believe that the Sembrani horse was the ancestor of the well-known Kuningan breed, to be dealt with below. Was this story, which may have circulated already much earlier throughout Western Java, also known to the Sultan of Banten? In that case he may have wanted a Sembrani horse in order to create his own breed.[21]

The Sultan sent his emissaries also to other places, such as Jepara, on central Java's north coast, where they asked the *bupati* (an indigenous functionary called *regent* [literally: he who rules] in Dutch sources) not only for horses, but also for cows and an elephant (1724/5). More often, his horse-buying envoys went to Cirebon, a sultanate to the east of Batavia (1711/12, 1715, 1725/26, 1729). The VOC headquarters in Batavia also acquired horses from Cirebon (1686, 1696, 1708). The area even seems to have had a reputation outside Java, as witness the arrival of horse-buying envoys from Palembang (Sumatra) and Siam (Thailand) during the first half of the eighteenth century.[22]

The first reference to Cirebon horses in Dutch sources dates from 1625. In 1680, one of the three Princes of Cirebon was said to have 200 horses at his disposal. Cirebon is one of the three areas that have produced breeds that in the nineteenth century were regarded as separate 'races', namely Priangan, Kuningan, and Kedu. Kuningan was part of the sultanate of Cirebon, a relatively densely populated, mountainous area in Cirebon's hinterland. So again we find a court that has 'outsourced' its horse breeding from the coastal plain to the fairly populous – but not yet too densely populated – upland valleys nearby, leading to the production and export of a surplus of horses not needed by the courts.

However, as early as 1817 the Cirebon horses were already in short supply. By the 1860s the Kuningan breed was said to have almost disappeared, probably because the Sultans had been removed from office in 1817 (thus removing royal patronage of horse breeding), and because the area had become too densely populated.[23]

Kuningan was just across the border from Sumedang, one of the Priangan Regencies. The first Dutch source to mention horses being bred in this area dates to 1686, and in 1694 Batavia acquired horses from Sumedang. It is possible that this was the beginning of a new breed of horses, the Priangan race, in the early nineteenth century one of the better known breeds. Apparently there were larger and smaller varieties to be found within this breed; the larger and more beautiful ones were the pride and joy of the Priangan aristocracy. There were several attempts to cross the original Priangan horses with larger, imported breeds, such as those from Persia and Arabia, which might explain their size. In the eighteenth century, horses were also to be found in other Priangan regencies, but those from Sumedang seem to have had the best reputation. Around 1820, the then *bupati* of Sumedang was an active horse breeder, crossing horses from Java with Makassar and Bima breeds.[24] During the early nineteenth century the Priangan breed, possibly an offshoot of the Kuningan horse, had arguably become the best known Javan race. In this case the 'court' of the VOC in Batavia had evidently stimulated the creation of the breed in the upland valleys of its hinterland by generating a constant demand. This had been supported by the Priangan nobility who relished deer hunting on horseback, and who were actively engaged in improving the Priangan race by crossbreeding it with European, Persian, Arab, Makassar and Bima horses.[25]

The Priangan horse may have been the first breed in the Archipelago to have been employed on a large scale for more mundane matters than royal pomp, carrying royal messages, and aristocratic pleasure (apart from the gold-transporting Minangkabau horses). At least around 1800 the more than 2,500 mares from Cianjur, one of the Priangan Regencies, were being used predominantly for coffee transport, because they had a greater carrying capacity and were faster than buffalo. But of course such activities were to be expected of what was possibly the only breed in the Archipelago to be regarded at least partly as a VOC creation.[26]

The Priangan breed seems to have been in decline by 1850, and by the beginning of the twentieth century it had become so rare that Groeneveld was inclined to regard all Java horses henceforth as one breed. This was no doubt the result of loss of pasture owing to increased population density, exacerbated by poor breeding practices. Moreover, the Priangan had undergone a political reorganisation in 1878, which put an end to the independent position of the Priangan aristocracy. Their hunting reserves were put to other uses, and their hunting days were over, which accelerated the demise of the Priangan horse breed.[27]

The VOC, although originally a merchant company, had become a terrestrial power in the Archipelago. Already in 1620 the fledgling establishment in Batavia felt the need for cavalry, ordering 100 saddles to be sent from the Netherlands. When they clashed with the Central Javanese kingdom of Mataram in 1628, they had not yet made much progress in this respect. While Mataram deployed 200 to 300 horsemen in the field, Batavia could only muster 24. As long as they were in conflict with Mataram, horses had to come from places further away, such as Gresik and Blambangan (eastern Java). In 1632, the VOC planned to import Persian horses to Ambon, in the Moluccas, and start breeding horses there, in order to be able to fight Mataram the next time on more equal terms regarding horses.[28]

From the 1660s onwards, relations between Batavia and Mataram improved to an extent. The Susuhunan and his *bupatis* often plied the VOC with horses, while the Company, in turn, sent Persian horses to Mataram. The VOC also started buying horses in areas under Mataram's sway, such as Cirebon and the Priangan (as was mentioned above), but also in Jepara, Mataram's main harbour (for Instance D 1667: 14 and 15 April; 8 and 11 June; 30 August).

During the Trunajaya War (1676-1681), relations between Mataram and the VOC changed dramatically, the VOC fighting one Javanese party as the ally of another. In 1679 Trunajaya had about 600 horsemen his disposal, while the VOC's allies had 400 to 500 cavalry. In 1678, VOC troops had captured 125 of Trunajaya's horses. This raises the question of the origin of these quantities of horses. In 1656, The Dutch emissary Rycklof van Goens had seen the Susuhunan's game park, where in addition to wild animals the ruler also let feral horses run wild.[29] However, for his armies, his jousting tournaments, and his establishment of service horses (*jaran gladhag*), the Susuhunan and his nobility needed a steady supply of animals, and it is questionable – given the quantities that may have been needed – whether they could rely on capturing and taming the animals from the game park.

From sources of a later date, we are familiar with the existence of a Javanese breed from Kedu. Kedu was one the most densely settled valleys in southern central Java. Up to 1812 it had been part of the territories of the Central Javanese courts, Yogyakarta and Surakarta, and before that of the undivided state of Mataram. Oral tradition, noted down c. 1840, suggested that around 1725 the Susuhunan of Mataram had started attempts in the Kedu area to get better horses by way of selective breeding of a race there already in existence. To this end he ordered the laying out of a fenced-in area (*krapyak*) in one of the upland regions of Kedu. Another nineteenth-century source states that Bima horses were used to improve the local breed.

Older data on the Kedu breed is sparse. In the 1840s it was regarded as the largest and most beautiful breed of the Archipelago. However, by the 1870s the race was said to have been in decline since Kedu was no longer under the authority of the Central Javanese courts, and one may assume that increased population density may have contributed to that.[30]

Kedu is an area where several interesting and mutually related developments occurred during the eighteenth, nineteenth, and early twentieth centuries. It was one of the most densely populated regions of Java, an extensive plain surrounded by volcanoes, characterised by rice paddies on the valley floor and the lower slopes of the mountains. The region was also the first in Java to grow smallholder tobacco on a commercial scale, beginning no later than 1746, and apparently of sufficient quality to be exported. Tobacco cannot be grown without fertiliser, and by the early twentieth century animal manure, including horse manure, was used here. It seems likely that this had been done since the earliest days of tobacco cultivation in the area, and it does not seem too far-fetched to assume that the presence of horses may have enabled the local smallholders to start growing tobacco. Both tobacco and horses seem to thrive in the cooler upland areas, as does maize, which could be used as fodder. However, two of the volcanoes in the Kedu area, the Sendoro and the Sumbing, had lost all their forests by 1850, a phenomenon doubtless linked to the large population of the area and to the demand for firewood (needed to cure tobacco), but perhaps also to the need for meadows for the horses.[31]

Central Java certainly already had many horses at earlier epochs. This is suggested by the previously cited VOC material but also by a much earlier source, namely Tomé Pires (1515). He distinguished the pagan hinterland from the Muslim coast. 'The Javanese heathen lords have richly caparisoned horses', he stated, referring to southern Central Java. 'They are great hunters and horsemen.' When he went to war with the coastal Muslims, the king had 2,000 horsemen (and, presumably, horses) at his disposal. It is possible that the coastal lords had fewer horses available. Pires only mentions the ruler of Tuban as the owner of many horses in addition to three elephants.[32] Could it be that Kedu was already then the main source of horses, and that he who controlled the inland areas was also in control of the supply of horses?

Eastern Java is difficult to interpret historiographically. Around 1515, Tomé Pires mentioned the lord of 'Gamda', to the east of Surabaya, who had many horsemen. More importantly, the ruler of Blambangan, who in fact held sway over most of eastern Java, 'has many horses in his lands; he alone has more than all the lords of Java, including Moors [Muslims]'. By Java he probably meant central and eastern Java only, but even then this is a fairly strong statement. Pires named the island of Madura, just off the eastern part of Java's north coast, as a place that had many horses too. Around 1780 Madura was again mentioned, together with Kedu, as horse country. It was even argued that the horses from eastern Java were larger then those of the west. Eastern Java was clearly well supplied with horses around 1800 (the first comparatively reliable statistics date from about 1820), and this was still the case a century later. Nevertheless, eastern Java does not seem to have produced its own breed in the eighteenth or nineteenth century, at least not one in the same league as those of the Priangan, Kuningan, or Kedu.

In the 1840s, one author named the eastern districts of Java as the area with the lowest quality horses of Java (with the exception of Bondowoso, in the Residency of Besuki). Another author also mentioned Besuki as good horse country around the same time. In 1860, Malang (Residency of Pasuruan) was reported to have good horses. In the early twentieth century, another author argued that in eastern Java the qualities of the Javan breed had been preserved better than elsewhere, because Europeans had tampered with it less than in the western parts of the island. And presumably, one would like to add, because at that time the amount of pasture available there was still sufficient. Nevertheless, it seems that these horse varieties never attained the status of a recognised breed.[33]

There are three factors that might explain why eastern Java does not seem to have developed one or more famous breeds (at least prior to 1850). In the first place there was a constant stream of imported horses from the islands to the east of Java. This was the case in 1515, when Bali, Lombok, Sumbawa, and Bima sold their horses (and their slaves) in Java, presumably mostly in places such as Blambangan, Surabaya, Jortan, Gresik, and Tuban. It was still the case at the beginning of the twentieth century, although by then Sumba, Savu, Roti, and Timor had supplanted Bali and Lombok. Horses from Makassar were also shipped to (eastern) Java.[34]

The second argument is that no court or local nobility seems to have attempted the improvement of local stock, as was the case with the western 'races' – although perhaps the princes of Madura, a minor court, may have been active in this respect. After the defeat of Blambangan by Mataram in the early seventeenth century, the area did not have a central authority, and during the eighteenth century a whole series of military campaigns turned eastern Java largely into a wasteland. And thirdly, we do not find the fairly (but not too) densely populated upland valleys here that seem to be a requirement, or at least a stimulus, for creating a popular breed.

Having dealt with the various regions of Java, what can be said about the island as a whole? In the first place it would appear that around 1500 horses had gained in importance since 1365. Although horses are already mentioned on inscriptions dating from the ninth century, they do not seem to have been very important around 1365. Horse riding is hardly mentioned in the Nagarakertagama, and neither are jousting tournaments. The 'meeting of the bamboo spears' to be found in a Chinese source dating from 1416 is a jousting tournament, but on foot![35]

Even then, the numbers reported by Pires are not really dramatic, if it is taken into account that Java must have had some four million inhabitants around that time. If it is assumed that based on the rough indications for Banten, southern Central Java, and Blambangan, there were 15,000 horses to be found on the entire island, a figure is arrived at of 4 horses per 1,000 inhabitants. This compares rather poorly to the c. 25 horses per 1,000 people in 1820. The 1515 figure for the whole of Java also pales in comparison to figures to be found for the Moghul court in or near Delhi (India) alone. Around 1590, Akbar was said to have 30,000 horses in Agra and Fatehpur Sikri (in addition to 12,000 elephants). In 1610, Jahangir also had 12,000 elephants, but the number of horses seems to have dropped in two decades from 30,000 to 12,000.[36]

The lower horse/people ratio in 1515 than in 1820 implies that the number of horses in Java had increased faster than did the number of people. This was no doubt partly caused by VOC demand, particularly from the late eighteenth century onwards. As we have seen, the VOC in Batavia required horses from Cirebon and the Priangan from the late seventeenth century onwards, actively stimulating horse breeding in these regions. Attempts were undertaken to improve the Priangan horses from 1736 if not earlier, and a stud farm established in Bandung by the Governor-General is mentioned in 1777. Around 1800, all Javanese 'regencies' along the north coast, from Tegal to Surabaya and perhaps farther east, also had to furnish horses for the VOC's *gladhag*.[37] With the construction of the infamous post-road from Batavia to eastern Java by Governor-General Daendels around 1810, a postal service with horses operating in relays had been created, which was also available for well-to-do travellers who could pay for it. Some European and Eurasian owners of private estates to the south of Batavia (Bataviase Ommelanden and Buitenzorg) are reported to have been horse breeders, or to have had meadows for the sole purpose of producing grass for horses.[38]

The trend of a continuously increasing ratio between horses and people stopped and reversed in the 1850s, as is shown in Table 3.1.

However, Java by no means produced all the horses it needed. This was so in 1515, when horses were shipped in from Sumatra, Bali, Bima and Sumbawa. It was still the case around 1800, when horses were being imported from Makassar, Sumbawa/Bima, Savu, Roti, and Timor, and around 1850 when Java received horses from Sumba and Australia. From the seventeenth century onwards Java also imported high class Persian and Arabian horses, including the crossbreeds from Jaffna. It could be said that Java had successfully 'outsourced' part of its supply of horses, thus leaving part of its 'ecological footprint' somewhere else.

Occasionally, Java also exported horses. The most remarkable importer during the period under consideration was Siam (Thailand). This seems to have started in 1666 and it continued at least until 1814. During the period 1680–1730 horse buyers employed by the King of Siam arrived almost every year in Batavia. They did not limit their shopping to Batavia and the surrounding area (where they probably bought Kuningan horses) but also went to Jepara, Semarang, and even to Mataram's capital, Kartasura, where they may have bought Kedu horses. They always bought between 40 and 60 horses, for which they received an advance from the VOC. It would appear that the prices slowly rose from around 30 Rixdollars (*Rijksdaalders*) for a horse in the 1680s to 40 in the 1720s.[39] This was rather expensive for a Java horse, and, as horse traders were rarely gullible folk, it may be assumed that they were purchasing higher quality specimens.[40] Nevertheless, if the

Table 3.1 Number of Horses per 1,000 People, Java, 1820–1900.

1820	1850	1900
25	40	15

VOC had not already established a regular Batavia–Siam trade link, this permanent flow of horses from Java to Siam would in all probability not have taken place.

Eastern Indonesia

Bali and Lombok were also occasionally mentioned as exporters of horses but this was already nothing more than a lingering memory by the 1850s and there is little other meaningful data. The literature is much more prolific on the races from Sumbawa and Bima (both on the island of Sumbawa), from Sumba (Sandalwood Island) and from Savu, Roti, and Timor. These islands share a dry climate and a sparse population, with large savannah-like areas where the horses could roam freely. All these islands had a number of rulers who often owned all livestock in their realms and who had slaves to take care of the animals. The breeds from Sumba and Sumbawa, in particular, had an excellent reputation. Savu and Roti sent their horses to Timor, and Dutch, Arab and Bugis ships picked up the animals from Timor, Sumba, and Sumbawa, transporting them to Java, but occasionally also to the Moluccas, Makassar, and areas outside the Archipelago.[41]

Bima and Sumbawa horses were mentioned as early as 1515, as was the horse trade from the island of Sumbawa to Java. In the seventeenth century the VOC imported small numbers of horses from Bima, but also from Dompo and Tambora, two other tiny sultanates on the island of Sumbawa. Sometimes the sources specify the colours of the horses, particularly when they are presented as gifts. 'Yellow', 'black' and 'white' horses are mentioned. Thomas Stamford Raffles, Lieutenant-Governor of Java from 1811–1815 and author of the well known *History of Java*, heaped praise on the Bima horses, which were said to strongly resemble Arabs in every respect except size. Around 1830 much larger numbers of horses were shipped from the island to Java, and these numbers continued to rise.[42] In the sources at my disposal, the other Lesser Sundas are mentioned much later than Sumbawa as breeders and exporters of horses. Horses in Timor are mentioned in 1699, but after that not until 1779, Savu horses in 1779, and those of Sumba and Roti not until 1830. However, by the latter date they seem to be already exporting large numbers of horses to Java, numbers that would increase as the century wore on.[43]

These areas differ in various respects from the 'western' regions where popular breeds had been reared. Population density here was very low instead of rather high, and the areas were almost certainly much dryer than the Sumatra and Java upland valleys discussed so far. However, the Batak and Minangkabau highlands as well as the Priangan, Kuningan and Kedu upland valleys were probably not as wet as the surrounding country, as they were located in the rain-shadow of the mountains. What remains is a clear difference in population density. It is possible that less labour is needed when the original vegetation already has a savannah character, and therefore burns much more easily than (wetter) forest vegetation. Apart from that, up to 1860 the rulers had slave labour at their disposal, and similar bonded arrangements under another name probably lived on after that date. What all areas had in common appears to be the presence of a state, albeit sometimes a very small one, that was interested in horse breeding. The advantage of the 'eastern', that is low-density, pattern is that horse breeding was not crowded out by the activities of an ever-increasing population of peasant cultivators.

It is not so easy to explain the widely differing dates at which horse breeding appears to have started on these islands, at least in so far as recorded by outsiders (Portuguese, Dutch, British). The Sumbawa breeds predate the European presence, and one surmises that Javanese influence transplanted an interest in horse breeding from Java to the island. Under the VOC a commercial link was forged early on, as the Company was interested in rice and sapan wood from Sumbawa. The political and economic links between Makassar and Sumbawa may have played a role as well. Although contacts between the VOC and Timor were also established quite early, here there was no pre-European tradition to fall back upon, and the trade link between Batavia and Timor may have been less intense, as no rice was involved, and the VOC was in fact almost exclusively interested in sandalwood. Finally there is Sumba, also a sandalwood-producing island (often called Sandalwood Island), hence the name sandalwood horses or sandalwoods for the horses of this place. In contrast to the rulers of Sumbawa and Timor, those of Sumba did their utmost to keep the VOC at bay, and no regular contacts were established until the beginning of the nineteenth century. It is not clear why they succumbed to the lure of money after 1800. The point here seems to be that under similar cultural, social and economic circumstances, a political decision – to keep the VOC at bay – may have postponed the advent of commercial horse breeding for half a century or more. However, market forces seem to have overcome the other considerations in the end.

Finally, with the Makassar breed, there is again a densely settled area, and rulers and noblemen with an interest in horse breeding. It is a rather dry area with savannah vegetation, thus representing a cross between the western (Sumatra, Java) and the eastern (Lesser Sundas) patterns. Horses were mentioned here from 1540 onwards. The region exported its horses to Java, and possibly to other places, at least since the seventeenth century.

The Makassar breed may have had its origin in imported Bima horses,[44] although the literature does not mention this.[45]

Conclusion

During the Early Modern period, the presence of a royal court and the availability of (upland) valleys with natural or human-induced grasses would appear to have been conducive to or perhaps even a requirement for the genesis of new horse breeds in the Indonesian Archipelago. In western Indonesia and in southern Sulawesi, these valleys were fairly densely – but not too densely – populated, while those to be found on the Lesser Sunda Islands were not. In the latter areas, slaves provided the labour that was necessary for the upkeep of the large herds of horses. In western Indonesia, slaves may have been involved too, at least in Sumatra, but in Java smallholders must have been responsible for most of the horse breeding activities. Although detailed evidence is lacking so far, it would appear that some of this labour (Kedu) was coerced, while in other areas (Priangan) horse breeding may have been undertaken by smallholders of their own free will as a commercial venture. However, some Priangan and Cirebon peasants should be regarded as serfs (although not fully-fledged slaves) and the possibility that the aristocracy of the region employed these people to look after their horses cannot be ruled out.

During the entire period under consideration, horses were traded from one end of the Indonesian Archipelago to another, while horses from other Asian areas were imported, and Indonesian horses were exported. During the seventeenth and eighteenth centuries, this system seems to have had two hubs – Batavia and the stud farm just off the coast of Ceylon. The numbers involved were probably not significant in relation to the total number of horses to be found in the Archipelago, but a constant stream of Arab, Persian, Makassar and Bima stallions must have influenced local breeds genetically, such as the ones from the Priangan and Kedu. Without the presence of a body such as the VOC, with its large ships that plied these waters on a regular basis carrying large cargoes, the trade in horses may have been much less important, on certain routes even non-existent, which implies that without the VOC some breeds, such as the Priangan one, might never have come into existence.

Much larger numbers of horses were involved in trade during the nineteenth and twentieth centuries than previously, particularly from Makassar and the Lesser Sundas to Java, as the latter island was no longer able to breed animals in sufficient quantities and of sufficient quality to meet local demand. This local demand also had a supra-local dimension as Java was producing more and more commodities for an international market and its population was growing faster than before – two phenomena linked to the presence of a colonial state.

From the mid-nineteenth century, the three Java breeds deteriorated in quality, partly owing to lack of pasture, partly perhaps owing to poor breeding practices, and partly because rulers and nobility had been deposed, or – if still in power – were no longer interested in (or capable of) maintaining them. After 1850, the number of horses per 1,000 people fell as well (Table 3.1), and after 1890 the total number of horses also went down. By then, steam power was driving out horse power. After 1900 the internal combustion engine would accentuate this trend, causing the number of horses to drop even further.

Early Modern Indonesia seems to have witnessed constant shifts from one horse producing area to another, which then, in turn, had to yield to yet another region. This is possibly what happened in northern Sumatra, when the centre of horse breeding moved from Aceh to the Batak upland valleys and from those to the Gayo area. A similar set of events seems to have happened in Java, and when the latter had to 'outsource' part of its horse breeding activities to the Lesser Sundas.

Generally speaking, horses were doing well in the Archipelago, and in Java it would appear that between 1500 and 1850 their numbers were increasing more rapidly than those of humans. As was shown in the case of Kedu, this may have had serious environmental consequences, such as deforestation.

Finally, an intriguing question is left unanswered. Is it plausible to assume that horse-raising was the reason that people were attracted to some areas in the first place? Is it possible to argue that some of the Batak areas, or some of the Priangan and Cirebon valleys became attractive to people because they could be used to breed horses? One thing appears to be clear, however: successful horse breeding in the more densely settled areas was not a sustainable activity. Perhaps its very success attracted so many people that the horses were displaced. However, it is also possible that population growth plain and simple, without any causal link to horse-raising, was the motor of this shift from one area to another.

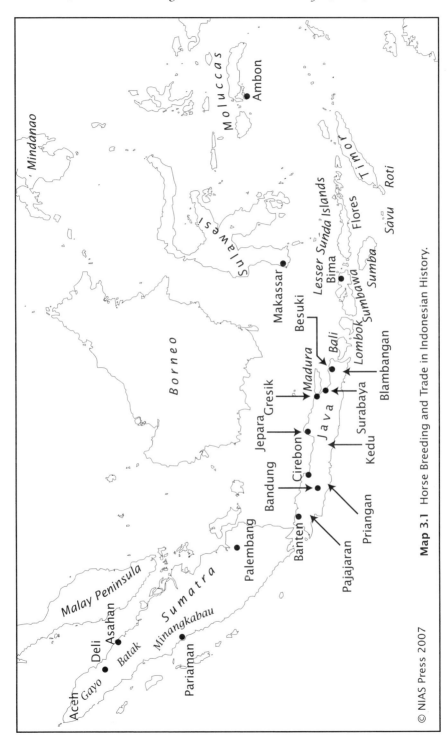

Map 3.1 Horse Breeding and Trade in Indonesian History.

© NIAS Press 2007

4. The 'Arab' of the Indonesian Archipelago: Famed Horse Breeds of Sumbawa

Bernice de Jong Boers

'Sumbawa produces the handsomest breeds of the whole archipelago; they are the Arab of the archipelago.'[1]

In the previous chapter, Peter Boomgaard explored the general traits of horse breeding in the history of several regions in the Indonesian archipelago.[2] This chapter specifically focuses on one of Indonesia's most famous horse breeding regions, the island of Sumbawa. It describes the different breeds as well as the methods of horse keeping that were employed there. It further shows the various roles horses played in the economy and cultural traditions of the island. Lastly, the effects of horse breeding on Sumbawa's natural environment are considered. But before all that, a short introduction to the geography of the island of Sumbawa is appropriate.

The Island of Sumbawa

The island of Sumbawa has an extraordinary rugged and indented shape with numerous bays. In the lower parts of this island, large grass plains alternate with brushes and trees; residues of forests can be found here as well. In these parts, mainly located on the north coast and around river basins, most agricultural activities take place and the majority of the population is concentrated here. The hilly uplands consist of forests and savannah. They are less populated and are used for *ladang*[3] cultivation.

Unlike many other islands in the Indonesian archipelago, Sumbawa does not have a tropical rainforest climate but a savannah climate with a distinct alternation between a dry season (from April to November) and a rainy season (from November until March). The dry period can be quite severe; often in this period Sumbawa looks very arid and barren. With a yearly rainfall of only 1,250 mm Sumbawa is one of the driest islands of Indonesia.[4]

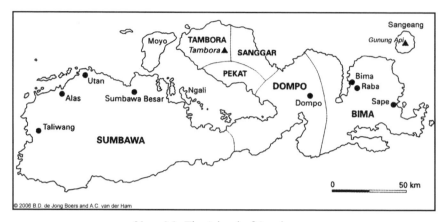

Map 4.1 The Island of Sumbawa.

Before 1800, human settlements were established throughout the island. All belonged to one of six princedoms namely Sumbawa, Bima, Dompo, Sanggar, Pekat and Tambora (see Map 4.1).[5] The principal export commodities of the island were rice, honey and beeswax, bird's nests, salt, sappan wood and horses.

The Introduction of Horses into the Archipelago and Sumbawa

Unfortunately, so far little is known about the introduction of the modern horse (*Equus caballus*) into Indonesia. Three quarters of a century ago, W. Groeneveld and J. C. Witjens stated that the Indonesian horse descended from a cross between Asiatic and African animals.[6] All the horses now known in Indonesia share characteristics of both these breeds. The authors further maintain that they were brought to Java by Chinese and Hindu merchants (from India) between 627 and 649 AD. Later, Arabs also brought horses from Southwest Asia.[7] In all likelihood, the horse was first introduced to Sumatra and only later to Java. After their introduction to Java, there was a slow equine diffusion to the other islands of Indonesia. But the total number of horses in the archipelago around the year 1200 AD was probably very small. It is even possible that the horse was not generally known in Java at the end of the fifteenth century.[8] What is certain, though, is that horses were introduced to the archipelago before the ninth century and are not often mentioned in sources dating from 1500 and earlier.[9]

In Sumbawa the domesticated horse was probably introduced by the Javanese in the fourteenth century (when the island was apparently a 'dependency' of Majapahit). At least this is what C. C. F. M. Le Roux concluded on the basis that in Sumbawa, both in the West-Sumbawan as well as the Bimanese languages, the horse is called *jaran*, the low-Javanese word for horse.[10] According to Nicolaus Adriani, the word *jaran* was actually derived from the Malay word *ajar* which means 'trained'.[11]

Various breeds are recognized from Sumbawa, usually simply named after their region of origin. Authors variously mention different breeds but no consensus exists in regard to this topic. A complicating factor is that most authors do not explain the criteria they use in distinguishing the various breeds. The Singapore Chronicle (1825) distinguishes two breeds: horses from Tambora and horses from Bima, the latter including horses from the tiny island of Sangeang (Gunung Api) along the northeast coast. William Thorn, a British Major, also considered the horses from Bima and Gunung Api as one breed but remarked that the finest horses were those obtained from the latter.[12] John Crawfurd also mentions the Tambora and Biman breeds but considers horses from Gunung Api as a separate and third breed.[13] W. H. Davenport Adams only mentions horses from Tambora and Bima, whereas G. W. Couperus distinguishes between Biman and Western Sumbawan horses.[14] Most authors do agree, however, that the Biman horse was the best breed. As Johannes Olivier remarked in 1828, 'They [Biman horses] are among the best races of Indian horses.'[15] By the twentieth century, authors tended to distinguish only two breeds: one from Bima and one from Western Sumbawa.[16] The horse from Tambora is not mentioned again for good reasons as the eruption of Mount Tambora in April 1815 wiped two of Sumbawa's princedoms, Tambora and Pekat, off the face of the earth, with none of their inhabitants surviving the cataclysm. In the other princedoms, too, numerous people were killed. Animal populations

Map 4.2 The Indonesian Archipelago.

were also devastated. As a consequence the Tambora horse breed became extinct.[17]

Many authors highly praised the quality of Sumbawan horses. They were generally described as small (their height at the withers seldom exceeded 1.25 metres), strongly built, intelligent and compliant. They were said to possess great strength and endurance.[18] F. H. H. Guillemard wrote for example: 'In spite of their small size [they] seem to carry almost any weight', while Heinrich Zollinger, a naturalist who spent three months on the island in 1847, concluded: 'The principal domestic animal of the island is the horse, which is found nowhere so much as in Bima and Western Sumbawa. That the breed is one of the best of the entire archipelago is well-known. Yes, many consider it the best.'[19] The most prevalent skin colours among Sumbawan horses were duns with a dorsal stripe, bays, greys and isabels (palomino). Blacks, chestnuts and piebald were rare.[20]

The earliest known source for horses on Sumbawa is the *Suma Oriental* of the Portuguese Tomé Pires in 1513. He noted that Bima had many horses which they exported to Java.[21] Another early source that mentions horses on Sumbawa is the book of the Portuguese Duarte Barbosa written in 1518. He writes that the people from *Cinboaba* (Sumbawa) and the island of *Oçape* (nowadays called Sangeang or Gunung Api, lying just off Sumbawa's north-eastern coast) 'travel on horseback and are good riders'. Moreover he calls them 'great cattle-breeders'.[22] So within 150 years after the presumed introduction of the horse into Sumbawa, the inhabitants of the island had already acquired great skill and competence with the animals – riding, breeding and exporting them. These early sources clearly show that horse breeding on the island antedates European influence.

Royal Horses at the Courts of Sumbawa

Nineteenth and twentieth century literature clearly establishes a link between the royal courts of Sumbawa and the breeding of, or at least the possession of horses. The rulers and aristocracy of Sumbawa mostly owned very large numbers of horses. Zollinger was told in 1847 that the Sultan of Bima owned 50,000 horses. He considered this an overestimation, however, and thought 10,000 horses a more accurate figure. Of this number, about 500 were stallions.[23] The population of Bima, both the nobility and the common people, were divided into hereditary groups called *dari*, which the Dutch sources often described as a kind of 'guild': 'because one of their characteristics was the specific nature of the statutory labor each of them was expected to perform'.[24] These *dari* provided specific tasks and services for the Sultan. One of these, the *dari Jara* consisted of the Sultan's grooms.

According to Zollinger, this *dari* comprised at least 100 men who were under the command of the *Jene Jara Asi* and the *Jene Jara Kapa*.[25] D. F. van Braam Morris in 1888 further mentions that they were not only grooms but had to guard and to train the horses of the Sultan as well as care for the maintenance of the saddles and girths. Van Braam Morris also makes mention of the *Dari Sangeang*, the inhabitants of Gunung Api, who were charged with the care of the Sultan's stud-farm on the island.[26]

A special feature of Bima since 1646 is the tradition of a 'holy horse' or a 'horse of Empire' known as the *Jara La Manggila*, which according to M. A. Bouman literally means: 'the horse that will not go slowly'.[27] The history of the holy horse had its origins on Gunung Api where a brown blazed horse with white stockings called *Kapitang* was owned by Sultan *Ambela Abil Khair Sirajuddin*. As the story goes, the Kingdom of Goa waged war against the Kingdom of Bone (both in Southern Sulawesi) in 1646. In this battle, the Sultan of Bima, who had married a Princess of Goa, lent assistance to his father-in-law mounted on *Kapitang*. Due to the horse's reputed prowess, the people of Goa could not be defeated. To commemorate the horse's meritorious behaviour, he was given the title 'horse of Empire'. Since then, every Sultan had his own 'horse of Empire' chosen from the offspring of *Kapitang* on Gunung Api.[28] This animal was not allowed to be mounted even by the Sultan.[29] Van Braam Morris writes that on special occasions, when the horse passed the government's fort, a salute was even fired. The horse was always preceded by a man bearing the royal lance and no one was allowed to cross the street in his path. Anyone who did so was fined one *real* by the horse's guards.[30] J. E. Jasper witnessed such a spectacle in 1908. After being bathed, the horse was brought to the market where rice, fruits and sweet potatoes were collected to feed it. What was left over was divided among the bystanders who accounted such foodstuffs a blessing that brought luck.[31]

Both the *Jene Jara* and the *Jara La Manggila* played a role in the ceremony called *Tuharlanti* conducted to install a successor to the throne as Sultan. Before the actual beginning of the inauguration ceremony the *Jene Jara* demonstrated their agility and equine skills. The horses, of course, were specially dressed for such an occasion with ornaments and decorations. Then the functionaries of the Sultan swore allegiance to him. During the ceremony, the Sultan was accompanied by the 'holy horse' which was also installed during a rite in which the *Jene Jara* paid homage to and took an oath of fealty to the animal.[32] Another occasion on which horses played an important role and one which is still practised today is the *U'a Pua* (or *Sirih Puan*) ceremony held on *Maulud*, the birthday of Mohammed. In this ceremony, several groups of horsemen perform manoeuvres and demonstrate martial arts on horseback.[33]

Apart from these ceremonies, horses also figure in Bima folk-stories and folk-beliefs. It was believed, for instance, that every five years a sea-horse visited the mares of the island of Gunung Api, which, after being serviced by him in his own element, brought forth 'the breed which is so beautiful and valuable'.[34] In another story, the horses of Bima play a more heroic role. The story tells how Prince Diponegoro famed for his rebellion against the Dutch in Central Java between 1825 and 1830 requested cavalry horses from the Sultan of Bima. They were dispatched to Java but arrived late, holding up the advance on the Dutch.[35]

Only the Sultan of Bima, the *Ruma Bicara* (Chief Minister) and the Dutch administrator (*Gezaghebber*) kept some of their horses in stables or on stud-farms, where the horses were provided with proper care and feeding.[36] This fostered their quality and, as a consequence, the trade-value of these horses. The ordinary people could not afford to build stables and simply let their horses freely wander in the pastures of Sumbawa. Van Braam Morris mentions that the Sultan's stud-farms were in the regions of Laambu, Kangga, Paie, Poja, Wera, Saie and on the island of Gunung Api. Moreover, J. P. Freijss noted that the Sultan of Bima even had his own boats for exporting horses to other islands, so avoiding transportation costs and making the whole enterprise all the more profitable. The returns were high and Van Braam Morris estimated that the Sultan's profits from horse trading amounted to about 1,000 guilders yearly.[37]

Less is known about horses at the court of Western Sumbawa. As in Bima, the Sultan of Western Sumbawa possessed a great many horses and just as in Bima the population were divided into a complicated system of hereditary groups, the *Tou Juran* and the *Tou Kamutar* that had to provide labour services for the Sultan. In Western Sumbawa, however, the nobility was excluded from this system.[38] The *Tou Juran* had to work on the rice fields of the Sultan. The *Tou Kamutar* were subdivided into small family groups called *ruwe* that resembled the *dari* in Bima. One of these *ruwe* was called *Tou rabowat aji*, literally meaning 'workpeople for the sovereign'. One group were obliged to cut grass for the Sultan's horses and another had to do the same for the horses of the *Dea Ranga* (Chief Minister). Albertus Ligtvoet also mentions that the Sultan of Western Sumbawa used the uninhabited island of Ngali, located in the Bay of Sumbawa, to graze his colts and let them develop into fine stallions.[39] According to Jasper, much the same use was made by the Sultan of Western Sumbawa of the island of Moyo.[40]

The Value of Horses

In a relatively inaccessible, hilly area such as Sumbawa, horses were indispensable for transporting goods (both as beasts of burden and draught animals) and people (as riding animals and pulling carriages – locally called *cidomo, Ben Hur* or *dokar*).[41] This was especially true at a time when the roads on Sumbawa were still very few and bad (roughly speaking this was the case until the second half of the twentieth century). Horses increased people's mobility and provided their main means of transportation. Many horse-tracks, therefore, could be found on Sumbawa, linking one village to another.[42] Gerrit Kuperus remarked: 'Only by using horses as riding and transport animals, is traffic possible between *kampung* [settlements] often located far from one another.'[43] Horses were also used in hunting, especially in deer hunts, as many Rusa deer (*Cervus timorensis*) were found on Sumbawa. Deer were hunted for their meat which formed an important additional item to the local diet especially after the introduction of Islam at the beginning of the seventeenth century when pigs became a forbidden food. Moreover, deer flesh in the form of *dendeng* (dried meat) was exported to Java and Celebes.[44] Finally, horses were also used in warfare but never for agriculture as was usual in Europe. On Sumbawa, only buffalos were used for this purpose. Horses, therefore, were of major economic importance in the local economy of Sumbawa. To the local people, they represented capital: horses were never eaten or reared as beef cattle (with just one exception to this rule, however, which will be mentioned below) as this incurred a loss of capital.[45]

Besides their use-value, horses also had vast trade-value. Important officials in the *Verenigde Oost-Indische Compagnie* (VOC, United East India Company) who wanted to monopolise the trade with Sumbawa were regularly presented with Sumbawan horses in return for the gifts of textiles and opium.[46] The Sultans and inhabitants of Sumbawa also obtained their weapons in this way. Horses were exchanged for guns, lead and gunpowder.[47] The Dutch soon discovered the fine quality of the horses and began to purchase them increasingly. From the *Overgekomen Brieven en Papieren* (or OBP in The National Archive, The Hague) it appears that Adriaan van Dalen was ordered to buy 20 to 30 horses in the year 1672.[48] He was further commanded to buy only those horses that were young and strong. The following year, 60 horses were purchased and a sum of about 22 rixdollars (*rijksdaalders*) for one horse was paid, a not inconsiderable amount of money at that time. Horses had already been exported to other islands by *perahu* (proa), especially to Java, Southern Sulawesi and to Southeast Borneo. The principal destinations on Java were Batavia and Surabaya where the horses were mainly used as draught animals. In Southern Sulawesi and Southeast Borneo, the horses were chiefly employed for hunting- and war-purposes. Sometimes, though, the horses of Sumbawa were transported further away. In 1717, the VOC transported horses from Bima to Ceylon where the company possessed its own stud-breeding facilities.[49]

Around 1800, the little realm of Tambora, situated on the peninsula of Sanggar had developed into an important horse-trade-centre.[50] Every year more than 1,000 animals were exported from this area. However, this situation came to an abrupt end when Tambora's volcano erupted with great violence. The results were disastrous for the whole island.[51] Some 10,000 inhabitants of the island died in the disaster as well as many horses. The whole island was blanketed in a thick covering of ash that stifled agricultural and pastoral pursuits. A great famine followed for men and beasts as well as an acute lack of drinking water that claimed many more human and horse lives. The economy of the island collapsed and the horse-trade came to a total standstill. The hardship was so severe that people were forced to seek alternative sources of nourishment and for the first time ate their horses in order to survive. This is the only time that the sources mention horse meat as foodstuff on Sumbawa.[52]

In the decades following the disaster, the situation recovered slowly. In 1821 Caspar Reinwardt paid a visit to Bima and observed that there were few horses and that only about one hundred were being exported.[53] After 1830, export of horses began again very cautiously but was limited to between one hundred and three hundred animals a year. Emanuel Francis visiting Bima in 1831 spoke of 'a fair amount of horses' and that in his opinion about 300 horses were exported from Bima to Java.[54] When Heinrich Zollinger arrived on the island in 1847, the horse population had recovered completely. He wrote that horses were seen there 'in an almost unbelievable multitude' and that anywhere one went on the island, one met horses gathered in enormous herds: in the meadows, in the mountains, in the valleys, in fallowed rice paddies and also in the most desolate wildernesses. Also the horse trade had got well under way again. Between 800 and 1,000 animals were annually exported to Bali, Lombok, Java, Bonthain, Singapore, Riouw and Malakka and brought a price of between 40 and 50 guilders each.[55] Every year ships even came from Mauritius and Bourbon to buy horses in Sumbawa.[56]

In 1854, J. P. Freijss estimated that Bima alone was exporting about 2,000 horses a year and that Western Sumbawa traded a further 3,000 or so. He noted the seasonality of the horse trade. During the east monsoon (from April until November) in particular, horses were transported by ships because the strong eastbound wind shortened the journey time to Java.[57] Prices for horses fluctuated between 20 and 50 guilders and the profits to be made attracted Arab traders to Sumbawa who had monopolised the trade by the end of the nineteenth century.

At the beginning of the twentieth century, the export of horses from Sumbawa reached its peak. At that time, the island possessed between 80,000 and 100,000 horses or between ten to fifteen per cent of the total horse population in the Netherlands East Indies and the number exported amounted to about 5,000 annually.[58] Most were exported to Java, mainly to Surabaya and Batavia but also to Semarang, Jember, Pasuruan and Besuki where they were used in the so-called *gerobak*.[59] Horses left Sumbawa through the harbours of Sumbawa Besar, Bima and Taliwang. Prior to 1910, horses were transported by local *perahu* or Arab schooner but between 1910 and 1920 the *Koninklijke Paketvaart Maatschappij* (KPM or Royal Mail Service) instituted a special *veelijn* (cattle line) between Java and the Lesser Sunda Islands.[60] Every fortnight, this steamer touched at Sumbawa Besar and Bima, loaded horses and transported them to Java. These cattle boats had stables on board and were staffed with *kulis* from the island of Lombok who brought the horses aboard.[61] Special loading-boats were used for this purpose and replaced the prior practice whereby horses had to swim out to the transport boats and were then hoisted aboard. During the passage to Java, the *kulis* stayed aboard to look after the horses.[62] As the capacity of these KPM steamers was much greater than that of the local *perahu*, more and more horses were exported by these boats. However, transport by local *perahu* continued and the competition between the two remained fierce.

It was at this point that the Dutch colonial government started to inter-fere in the island's horse trade. They were convinced that Arab middlemen and their suppliers on Sumbawa were only interested in short-term profits and selected only the best horses which they sold as expensively as possible. The long-term consequences of this policy, the Dutch thought, would result in only inferior horses being left on the island and to a deterioration in the quality of the stock.[63] The colonial government, therefore, took a number of measures to improve the breed.[64] A veterinary service to combat veterinary diseases was founded in 1912 and located in Sumbawa Besar, while at the same time a number of Arab and Australian stallions were purchased to cover with local mares and improve the stock. This programme ended in complete failure as the ensuing offspring were not well proportioned.[65]

An alternative improvement programme followed quickly. This was called the *reinteelt* programme (selective breeding). The measures taken within this programme implied that from now on a careful selection of stallions and mares needed to take place and that inferior stallions had to be castrated.[66] Moreover, horses had to be meticulously registered and export-limits had to be formulated. These measures proved very unpopular among the inhabitants of the island. On Java and the rest of the archipelago as elsewhere, geldings and mares were considered to possess less strength than stallions. For the local people of Sumbawa this meant that mares and geldings had little commercial value. Stallions fetched the best prices and the most lucrative profits were to be made from selling these first. Such sentiments may explain why the *reinteelt* programme did not have the intended effects.

In the meantime, the horse trading continued as usual. During this period, Sumbawa had to cope with tremendous competition from the other equine-exporting islands such as Sawu, Timor and Sumba. In particular, rivalry with the famed Sandalwood Horse (from Sumba) was great. However, the horses from Sumbawa were cheaper and the island lay in a more favourable location for transportation to Java and Sulawesi than did Sumba. These factors help explain how Sumbawa maintained a competitive position and remained the largest exporter of horses in the archipelago.[67]

From the 1920s, motorised traffic appeared on the scene. The role that draft and pack horses had formerly played began to decline. The demand for horses diminished and prices fell. Some people even predicted that the horse trade would come to a definitive end.[68] However, this was not the case. Although the number of horses on Sumbawa gradually decreased, its population today at about 53,000 animals can still be called substantial. Moreover, the inter-island trade in horses continues.[69] The high cost of motorised vehicles ensures that a certain demand for horses persists. Besides that, the use-value of horses remains high on Sumbawa. In out-of-the-way, hilly regions of the island, horses are still indispensable for the transportation of both people and goods. Moreover, many *Ben Hurs* can still be seen along the streets of the regional capitals of Sumbawa Besar and Bima despite the great competition of minibuses.

Horse-Keeping Methods and the Environment

Two methods of keeping horses existed on Sumbawa. The first involved keeping one or two horses around the house for daily use, especially as draft or pack animals. The second method involved breeding horses. This horse-breeding was not very labour-intensive. Horses in so called couples (one stallion and a number of mares) were left to wander around freely, grazing on the plains, in fallow fields and harvested paddies where they enjoyed complete reproductive freedom. As the horses fed themselves, little time was required to take care of them. At most, cultivated lands had to be fenced off. This method of horsebreeding was called 'semi-wild' breeding because sometimes the animals reverted back to their former 'wild' status.[70]

This labour-extensive method of horse-breeding was particularly well adapted to conditions on Sumbawa as the island was relatively sparsely inhabited. Its population was estimated at between 170,000 to 300,000 people prior to 1815, resulting in a low average of about 11–19 persons per square kilometre.[71] As a result labour was scarce but there was land aplenty.

Therefore, the principal form of agriculture on the island, *ladang,* was also consequently not labour-intensive. With *ladang* agriculture, the original vegetation of the land is felled and set on fire, after which it is cultivated for one or two years and then allowed to remain fallow for a long period. This method of agriculture perfectly complemented extensive methods of cattle-breeding. Moreover, the advantage of this manner of labour-extensive cattle-breeding was that it required only a small amount of money and so horses could be produced inexpensively.

A great part of Sumbawa's natural landscape consists of open forests, savannah and extensive grass plains. Over and above these areas, much of the agricultural land on the island is only used temporarily. Together with the natural vegetation of the island these semi-abandoned agricultural lands, like *ladang* lying fallow and rain-dependent *sawah* lying idle during the dry season, form excellent pasturelands for livestock.[72] The animals simultaneously fertilise the soil through their droppings which improves yields in the next farming season.[73] As to fauna, there are no large animals of prey that might threaten the horse population.[74] Moreover, the soils are rich in lime, which is important for equine reproduction.[75] All these factors contributed to making Sumbawa suitable for successful horse-breeding. As most of the island is unsuited to agricultural activities, livestock-keeping is a good way to make these lands economically productive as well as being an ecologically sound adaptation to environmental conditions.[76]

Still, environmental factors were not all favourable for horse-breeding. The dry season adversely affected horse-breeding. During this season it was often very difficult for horses to find food. They had to traverse long distances just to find drinking water and edible grasses with the consequence that animals often became emaciated and depleted. Over the lengthy dry period, green pastures withered to yellow-brown, became scarce to find or even disappeared. Old, weak and sick animals often did not survive these kinds of circumstances.[77] Sometimes precautionary measures were taken to prevent too high a mortality. Before the end of the dry season, the *alang-alang* (*Imperata cylindrica*) was set alight to allow new, green and edible grasses to replace the old and less edible ones.[78] Other measures taken at the end of the nineteenth century (and perhaps earlier) to combat the dry season involved providing water troughs, supplemental feeding and foliage, and relocating horses to higher, more wooded areas.[79] In Raba (a place in the neighbourhood of Bima), rice and straw (made from the peanut plant) were sometimes stored for times of distress. On other occasions, corn was planted in the dry season specifically to be fed to animals.[80] Nevertheless, hundreds of horses died yearly from lack of food and water according to the veterinarian Ajoebar stationed on the island.[81] Other threats to horses were endemic diseases like glanders, anthrax, hemorrhagic septicaemia (bacterial blood-poisoning) and surra (trypanosomiases).[82] Glanders and anthrax were highly contagious and caused recurrent and violent epidemics among horses with high mortality rates before the twentieth century.[83] Alternatively, some authors contended that the horses were hardened by these kinds of difficult circumstances and that this was the very reason why Sumbawan horses had grown into such powerful and muscular beasts renowned for their great perseverance.[84]

It is beyond dispute that the landscape of Sumbawa changed considerably with the introduction of horses. Since a full-grown horse requires extensive grazing, the need for pasturelands must have increased enormously. This expansion can only have been at the expense of woodlands and was compounded by the regular firing of grasslands to provide edible young grass as fodder. Such fires, once lit, could also get out of hand and burn down larger wooded areas than were intended. The consequence of these fires might not even necessarily be seen as calamitous by the local population: grasslands were appreciated because they attracted deer. Grasslands not only constituted good pasturelands but were also excellent places for deer-hunting, a beloved local leisure activity. Moreover, the presence of horses in fallow *ladang* fields left young saplings with far less chance to regenerate and for the forest to recover, eventually leading to a further decrease in woodlands over time.

Initially over-grazing was not a problem and only occurred when the number of animals per square kilometre was such that the natural vegetation was eaten away faster than it could regenerate, resulting in serious erosion. Since the population of Sumbawa was low (still only an average density of 21.4 per square kilometre as late as 1930) and the agricultural area was limited, there were sufficient grasslands.

In the 1920s and 1930s, Sumbawa owned about 100,000 horses or between 10–15 per cent of the total population in the Netherlands East Indies, a high number for a relatively small island.[85] Compared to other islands in the archipelago, Sumbawa was a real 'livestock-island'. In 1933, it had 601 animals per 1,000 inhabitants, a rate higher than Madura with its 392 animals per 1,000 inhabitants and four times more than the average rate for the whole colony at 149 animals per 1,000 inhabitants.[86] The number of animals on Sumbawa corresponds to an average livestock density of 12.2 animals per square kilometre, the highest in all the Lesser Sunda Islands. While Madura had the highest number of livestock in the archipelago (140.5 animals per square kilometre), Sumbawa had the highest horse-density with six animals per square kilometre.[87]

All in all the introduction of horses to Sumbawa is an interesting example of how animals change a natural environment. Although these changes were considerable, a crisis was avoided. As the density of the population was low, the carrying capacity of the natural environment was resilient enough to support the considerable number of horses. By the time population pressures began to mount on Sumbawa after 1930, the high tide of horse trading was already over and the horse population had begun to diminish.

Concluding Remarks

Sumbawa has been a major horse-breeding and horse-exporting island since the sixteenth century and horses have constituted one of the island's principal commodities. Horse-breeding as practised on Sumbawa clearly makes sense from both an ecological and an economic point of view. There was a suitable natural environment: the island had many natural grass lands, good limy soils and a lack of large prey animals. Moreover, horse-breeding could easily be combined with other economic activities like swidden and (rain-dependent) *sawah* cultivation. As a result of the relatively small human population and the ample space available, the method of horse-breeding on Sumbawa was very labour-extensive.

Thus horse-breeding was suitable for Sumbawa but was Sumbawa suitable for horses? Looking from a horse's perspective, it seems very unlikely that any horse would ever have chosen Sumbawa as a preferred location. Horses were hardly looked after and they suffered and often perished as a consequence of the yearly drought during which food and water were hard to find. Natural hazards and various diseases like hemorrhagic septicaemia, surra, glanders and anthrax constantly menaced their lives.

Having horses was a real economic asset for the local population for centuries; they were indispensable in daily life as beasts of burden and as a means of transport. Moreover, they fulfilled an important role in war and hunting activities. Apart from their use-value, horses also had a profitable exchange-value. They could serve as gifts or were used as goods in barter and trade. The trade in horses was especially profitable to the higher echelons of the society: the sultans, the nobility and the Arab merchants. Finally, and related to this, horses were symbols of pride and status and fulfilled important roles in court life and rituals. During the 1920s and 1930s, with the advent of motorised traffic, the horse almost completely lost its commercial value. As a consequence, their numbers declined during the second half of the twentieth century to a current population of around 50,000. Today, an inter-insular trade continues but on a much smaller scale than in 1900. Having largely lost their commercial attraction, the fame of the Sumbawan horse breeds has almost entirely faded away outside the island.

Locally, however, horses are still very much appreciated. To commemorate their glorious past and as a symbol of their still substantial use-value, statues of horses have been erected in both the regional capitals of Bima and Sumbawa Besar. Despite the competition of minibuses, horse-carts remain a very important means of transportation on the island. Therefore, horses remain a prominent feature of Sumbawa's present-day landscape, almost as if nothing had changed since the sixteenth century.

5 Javanese Horses for the Court of Ayutthaya

Dhiravat na Pombejra

Introduction

Along with elephants, *naga*, and other beasts, horses featured in stories of the Buddha, and thus also in Siamese art and illustrated manuscripts.[1] In the *Tripitaka* the Buddhist universal monarch or cakravartin had a horse (along with an elephant) as one of his seven accoutrements.[2] For centuries, these animals played a role, too, in the warfare of continental Southeast Asia.

During the seventeenth and eighteenth centuries horses – and Javanese horses in particular – became trading commodities much sought after by the royal court of Siam. Contemporaneous Dutch documents both from Ayutthaya and in Batavia mentioned the horse-buying expeditions sent out by various Siamese kings. This chapter is a preliminary study of how this trade was conducted by the kings' men, with the sometimes reluctant cooperation of the Dutch United East India Company (VOC).

During the reign of King Narai (1656-1688), the court of Ayutthaya started buying horses from Java. During the two years of 1691 and 1692 alone, horse-buyers of King Phetracha (1688-1703) bought a total of 73 horses from the interior of Java.[3] By 1725, the court of King Thaisa (1709-1733) was still buying horses from Java, purchasing 53 horses in that year alone.[4] The VOC records contain many such references to Siamese horse-buying expeditions from the 1680s till at least the 1730s, during the reign of King Borommakot (1733-1758).[5] The question remains, though, why did the Siamese seek to buy Javanese horses? The 'noblest' specimens of horse known to the Asian courts were after all Persian or Arab (Arabian). In all likelihood it must have been easier, and presumably cheaper, to buy Javanese horses. It is not clear what kind of horse is meant by the term 'Javanese horse' in the VOC sources. According to the Portuguese writer Tomé Pires, writing in the early sixteenth century, Sumatran horses were shipped regularly from Eastern Sumatra to Western Java, while horses from Java were in turn shipped to the islands of Bali, Lombok, and Sumbawa.[6]

Map: Kreangkrai Kirdsiri

Map 5.1 Ayutthaya and Java.

The Java pony as we know it today has supposedly been 'decisively influenced' by the Arabian breed. According to the *Encyclopaedie van Nederlandsch-Indië* the Javanese horse, in the seventeenth century, was 'een sterk en deugdzaam paard' (a strong and virtuous horse) that was still giving sterling service during the Java War of 1825–1830. By the early twentieth century (when the Encyclopaedia was published) the breed had 'degenerated', though it was declared that fine specimens could still be found in Besoeki and Preanger.[7] Nowadays the Java pony is used primarily as a pack or work animal. Breeds similar to the Java pony are to be found on other Indonesian islands, such as Sumatra, Timor, and Sumba (see Chapter 4 in this volume). According to a mid-seventeenth century Dutch source (De Graaff), troops of wild horses could still be seen roaming the interior of Sumatra.[8] As for the Timor pony, it is of a more delicate build (and smaller) than the Javanese, while the Sumba is used as a dancing horse.[9]

It is not surprising that the Javanese horse has been influenced by the Arabian and Persian breeds. Commercial and cultural contacts, including religious influences, had been key elements of the relations between the Middle East, Muslim India, and island Southeast Asia for centuries. The horses brought as gifts to local rulers inspired them to search for more.[10] The importation of Arabian or Persian horses almost certainly led to cross-breeding with local ponies.

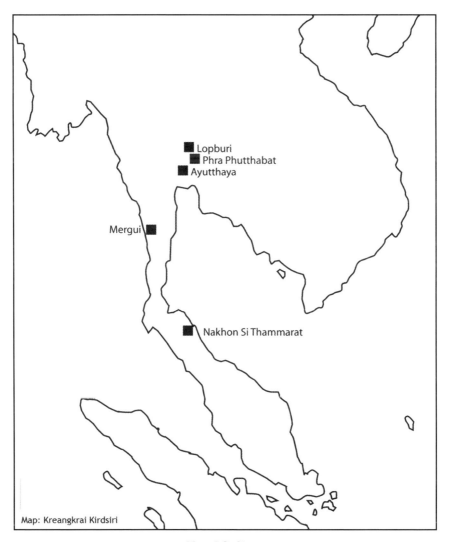

Map 5.2 Siam.

The Siamese Context

The area near Ayutthaya, the old capital of Siam, with its annual inundation during the rainy season, was not very suitable for horses. The riverine, amphibious way of life of the Siamese Central Plains meant that the primary means of transportation was by boat. John Crawfurd, in his *Descriptive Dictionary of the Indian Islands & Adjacent Countries*, claimed that the Siamese horse, 'like that of all the countries south-east of India as far as China, is a pony, not exceeding 13 hands high, and is in general use only in the uplands, being rarely seen within the tract of inundation.'[11] To the Siamese of Ayutthaya, however, the horse had its significance and usefulness.

The horse, along with the elephant, was highly regarded by the monarch and the royal court. It is thus not surprising that it should have formed part of armies and royal processions during the Ayutthaya period. Evidence on horses before 1600 is scant, but for the seventeenth century, there are western sources which mention horses in the context of the Siamese king and his court. It is clear that royal processions involved the use of horses as well as elephants. The Dutch merchant Joost Schouten refers to the king's horse guards forming part of the royal procession by land.[12] Schouten's successor Jeremias van Vliet, in his *Description* of Siam, describes the land procession in even greater detail, saying that the horses and elephants of the king were 'adorned copiously with gold and precious stones'. He relates further that the people riding the horses in the royal procession were 'courtiers and great men', meaning the *khunnang* or Siamese officials.[13] This suggests that the basic skills of horsemanship were probably mandatory for a young nobleman of the seventeenth century.

According to the French missionary Nicolas Gervaise, when the king 'goes out in order to display his majesty in the presence of the ambassador of some ruler to whom he personally wishes to do honour, or in order to attend the ceremonies of some great festival, the companies of his footguards are supported by troops of cavalry consisting in all of between twelve and thirteen thousand men'. Another Frenchman, the diplomatic envoy Simon de La Loubère, refers to the King of Siam's employment of 'a foreign standing Horse-guard, which consists in an Hundred and Thirty Gentlemen' of Mughal 'Moors', plus some additional horse guards of 'Meens' (probably Khmers, in Thai *khamen*) and Laos. The Mughal horse guards must have been elite troops of the king, probably first employed during the period of 'Moor' ascendancy at King Narai's court. The king's horse guards, in common with other foreign mercenaries, resided outside the Royal Palace but were called upon to accompany the king when he went out.[14]

In recounting the 1685 French embassy's passage through the courtyards of the Royal Palace in Ayutthaya, the Abbé de Choisy mentions the king's horses. The second courtyard at the palace contained 'a squadron of perhaps 300 horses'. Even to a seventeenth century Frenchman like Choisy, the horses of the King of Siam seemed 'rather fine', though 'badly trained'.[15]

A plan of the Royal Palace at Ayutthaya drawn by the German physician Engelbert Kaempfer shows '[a] large place for the running of races'.[16] It is not known whether these races were horse races. A notable example of horse-racing in Southeast Asia would be that in Aceh.[17] The ponds mentioned by Kaempfer were probably used for washing and bathing the animals.

Were horses used in tournaments, or in ceremonies other than the royal processions? In late eighteenth century Java, the court of Sultan Mangkubumi was much taken up with feasts, which included tournaments by horsemen, along with animal fights.[18] Some evidence on Siam is also available: in April 1639, during major festivities at King Prasatthong's court, horse-riding was, along with fights involving elephants, part of the entertainment on offer. The Palatine Law of Siam, dating from the fifteenth century, mentions playing polo (*khli*) as one of the kings' activities.[19]

Jeremias van Vliet describes the army of King Prasatthong in some detail. He claims that the king had over three thousand elephants, with 400–500 alone in and around Ayutthaya, each with two to three men attending them regularly. According to Van Vliet, '[t]he army also possesses ponies but no special horsemen are provided for. The cavalry are armed with old muskets and leather shields, so that an army provided with modern weapons does not need to fear an attack of the Siamese cavalry.'[20] It is noteworthy that Van Vliet uses the term 'ponies' to describe the horses in the Siamese cavalry. In La Loubère's opinion, too, horses were not much used by the Siamese in warfare: 'As they have no Horses (for what is two thousand Horse at most, which 'tis reported that the King of *Siam* keeps?) their Armies consist only in Elephants, and in Infantry, naked and ill-armed, after the mode of the Country.'[21]

Not long before the French involvement in Siam, however, a Dutch document records that an Ayutthayan expedition sent by King Prasatthong to subdue Songkhla was 25,000 strong, with 300 elephants and 'many horses'.[22] It would be fair to conclude, however, that elephants were the 'elite' cavalry of the armies of Ayutthaya, while the horses were very much of secondary importance. The decisive duels between kings and generals were conducted on elephant-back, after all. Horses were useful in that they were more mobile than elephants, and certainly faster. Messengers and scouts were those who could make best use of the horses.

Ibn Muhammad Ibrahim Muhammad Rabi, author of *The Ship of Sulaiman*, the account of the 1685 Persian embassy to Siam, recounts stories about

the Siamese attitude to horses. First of all he relates that the members of
that embassy had brought fine quality Arabian- and Indian-bred horses
with them, and it would have been very burdensome to transport them
back to Persia. 'Of necessity these horses became gifts to the Siamese king's
estates.' The Siamese apparently had no interest in these 'fairy-like beaut-
ies', and although they knew the real market value of the horses they paid
low prices for them (presumably a prerogative of royal 'monopolism'). The
Persian source goes on to describe briefly the Siamese way of training their
horses, which was to teach them to lower their heads in deference to their
riders. Also, Siamese cut the manes and tails of their horses because they
considered it bad luck for both rider and horse for the animals to have long
manes and tails.[23]

The royal chronicles of Ayutthaya, too, would seem to offer plenty of
evidence for the use of horses in traditional Siamese warfare. From the epi-
sodes dealing with the reign of King Naresuan alone, for instance, there are
data which says that horses were used in reconnaissance and news-bearing,
as well as in battle. During a war against the Burmese, King Naresuan advised
his men at Martaban to station 20–30 fast horses so that they might take
turns to take news to the main army. Horses were used in greater numbers
than elephants, even, in the army assembled to invade Cambodia. That force
comprised 100,000 men, 800 elephants, and 1,500 horses.[24] The Burmese
armies had horses as well as elephants in their cavalry. A key part of King
Alaungpaya's armies which invaded Siam in 1760 were the Manipuri horsemen, who later formed the rearguard as the invasion forces withdrew.[25]

Siamese sources other than the royal chronicles contain plenty of references to horses too. While the chronicles mostly refer to horses in the context of
military campaigns, the Testimony of Khunluang Wat Pradu Songtham testifies
to the existence of stables for the king's horses on the walled island-city of
Ayutthaya.[26] A Dutch document concerning the succession dispute of 1703 which
to all intents and purposes was testimony given by a Siamese courtier at the
court of King Sua, recounts how the king and his half-brother Chao Phra Khwan
rode horses to the funeral of their father King Phetracha. Chao Phra Khwan was
encouraged to ride a horse, being given 'the best horse from the king's stables',
and was later lured to his death while riding in the vast grounds of the Royal
Palace. A group of his half-brother's most trusted courtiers dragged him down
from his mount and 'executed' him with sandalwood clubs. The prince's killing
left the way clear for Phrachao Sua to enjoy his rule as king. The whole episode
shows that Siamese court culture, at by the end of the seventeenth century, had
become to some extent a 'horse culture'.[27]

There are old Siamese horse-riding and horse-identification manuals
still extant, though these probably date from the early to mid nineteenth

century rather than the Ayutthaya period proper. But at least in the matter of the valuation of horses, the standard was exactly the same during the late Ayutthaya and the early Bangkok periods. The Siamese in olden days seemed to attach a very great significance to the colour of the horses. The illustrations in the old manuals clearly differentiate between the colours of various 'types' of horses, rather than on their physical or structural characteristics. Much importance is also given to the classification, in a strict order corresponding roughly with social hierarchy, of various kinds of horses. They do not, however, allude to the provenance of horses. On the matter of the horses' colour, the manual states that a horse which has a white body and a black head was deemed to be fit for a king, along with a black horse which had a white tail, or all-white and all-black horses. No explanation for these preferences is given by the manuals. A secondary class of horse could nevertheless have sterling qualities of endurance and courage too. Less extraordinary specimens were classified as being of the third and fourth class.[28]

The illustrated Siamese manuscripts depicting seventeenth century royal processions by land and by water, thought to be copies of the Wat Yom ordination hall murals, show few horses. Curiously enough, horses appear only as part of the water or royal barge procession, twenty horses progressing on land beside the more illustrious of the barges.[29] The emphasis of the land procession was on elephants and infantry, and no horses are depicted, contrary to data in seventeenth century western sources such as Schouten and Van Vliet. A royal procession at the Phra Phutthabat (Buddha's Footprint) shrine in 1737 was said to have included four Persian horses, and other horses mounted by the king's retinue. King Borommakot himself, however, was on elephant-back.[30]

The Siamese had always used horses for transportation, warfare,[31] and ceremonial purposes, long before the coming of either the Persians or the Europeans to Siam. But where did Siam's horses come from, since there seems to have been no native breed? For the earlier, pre-1600 period, there has been no detailed research, and it is difficult to know where to find data for such a study. But for the seventeenth and eighteenth centuries, the VOC sources may provide some illumination. The Dutch data come from three main sets of archival materials. The most informative and detailed sources are the letters and reports of the VOC merchants stationed in Ayutthaya and the letters of the *phrakhlang* minister (in the *Overgekomen Brieven en Papieren* series). The *Generale Missiven* or general letters of the Governors-General and Council of the Indies, and the *Batavias Uitgaand Briefboek* or letters from Batavia to the various VOC offices in Asia, are also very useful sources in the study of the trade in Javanese horses.

Buying Horses for King Narai (1656–1688)

The earliest Dutch sources to mention the buying of horses for the Ayutthaya court date from the early 1680s, during the reign of King Narai. However, an earlier source states that in 1651 the VOC Governor-General Carel Reniers sent 'two Javanese horses' as gifts to King Prasatthong's favourite Okya Sombatthiban.[32] Around 1681–1682 the King of Siam (Narai) sent a vessel containing a party of horse-buyers to purchase horses in the Cirebon area of northern Java. According to Chaophraya Phrakhlang (Kosa Lek)'s letter, the Siamese ship was on its way from Batavia to Cirebon when it was attacked and burnt by the Javanese. The court therefore asked Batavia to look into the problem.[33] The Siamese must have started to become increasingly wary of sending their own vessels to buy the horses, because they came to rely more and more on the Dutch.

The area around the *pasisir* towns on the 'East' or north coast of Java seems to have been a popular place to buy horses. In 1683–1684, the Sultan of Jambi asked the VOC for passes to enable his men to go to buy horses on the East Coast of Java. The sultan wanted 40–50 riding horses for use in the 'tournoybaan'. In 1685 the Sultan also purchased some 50 horses, though this time his men were said to have gone to the island of Bangka off the Sumatran coast.[34] In 1715 the Sultan of Banten was still interested in buying horses from areas further east on Java. He sent his brother and uncle to Cirebon, on the East Coast of Java, to buy horses during that year.[35] As late as 1726 a 'Kjahi Astradipana' from Palembang went to Cirebon to buy horses.[36]

It is not mentioned in any contemporary sources why King Narai wanted to buy Javanese horses. A couple of possible clues may be that the king employed 'Moors' ('Mughals' and Persians) as his horse-guard, and that in the reign of King Phetracha (1688–1703) horses were required for the king's 'cuirassiers'. Perhaps King Narai's horse-guards wanted new horses on a regular basis, or wanted to improve the quality of the king's horses through crossbreeding with superior specimens. According to La Loubère, King Narai already had 'a dozen of *Persian*', gifts from the King of Persia which by 1687 had already depreciated in value. The Siamese king '[o]rdinarily ... sends to buy some Horses at *Batavia*, where they are all small and very brisk, but as resty as the *Javan* people are mutinous.' When La Loubère stopped by in Batavia on his way to Siam, he found two Siamese there 'to buy two hundred Horses for the King their Master, about a hundred and fifty of which they had already sent away for *Siam*.'[37]

A letter from the *phrakhlang* minister to the Governor-General and Council dated Chulasakkarat 1045/6 (1683) asks Batavia to supply the

Siamese court with thirty stallions and thirty mares. The minister explains that the court did not send people to buy horses in Java anymore, because it caused the VOC to incur losses. This was possibly a piqued Siamese reaction to VOC complaints about expenses. But it did not prevent the Okphra Kosathibodi in 1684 (presumably the acting *phrakhlang* following the death of Kosa Lek in 1683) from asking for the Governor-General's cooperation in helping a group of the king's men previously sent to fetch sixty horses from Java.[38] However risky the enterprise was, the Siamese were at this stage still occasionally sending their own ships to buy and transport the horses, probably in order to be less reliant on the Dutch.

There had been two junks sent to transport horses from Java to Siam with the Siamese embassy to Batavia of 1685–1686. A 'Gravaminas' or list of grievances presented to the Governor-General and Council in Batavia by the Siamese ambassadors shows that sending ships and buyers to Java to procure horses for the king was something that happened on a regular basis. But Governor-General Van Goens had earlier written to the *phrakhlang* requesting that the Siamese court desist from sending ships to buy horses in Java: the Siamese were coming into direct competition with the VOC by selling textiles in Java, and were anyway making losses on their voyages. Van Goens offered to help King Narai procure whatever horses he wanted from Java. The Siamese court, therefore, asked that the Company send thirty stallions and thirty mares to Ayutthaya. According to a VOC general letter of May 1684, these sixty horses had not been obtainable.[39] It was perhaps very difficult to obtain mares from the parts of Java controlled by the court of Mataram.

The Siamese court was nothing if not persistent in its pursuit of goods desired by the king. This perseverance eventually brought about positive results. There is no record of whether horses were taken back to Siam on the king's ships in 1685–1686, but a letter dated December 1686 mentions about 39 horses being sent to Siam per the *Walstroom* on 23 May of that year. The VOC ship also took back to Siam the Siamese king's envoy Okluang Chula, who had bought these horses while discharging his diplomatic duties in Batavia.[40]

The following year the Siamese court and the VOC arrived at an arrangement which was to prove the model for later horse-buying activities. The VOC authorities in Batavia advanced money to the king's horse-buyers (probably the ones La Loubère encountered), and also transported the horses to Siam. In 1687 the sum of money advanced was 1,609 rials. During that year 67 horses bought by the king's men were sent on VOC vessels to Siam.[41] In this way the Siamese court obtained the requisite number of Javanese horses, while at the same time the Dutch did not have to worry about any Siamese competition in bringing textiles to sell in Java. But even at the very end of King Narai's reign, the shipping arrangement was still flexible and non-VOC vessels could also be used to ship the horses to Ayutthaya. In 1688, the final year of the reign, a VOC general letter mentions that although 24 horses for the King of Siam had been sent with the VOC vessel *Lek*, there was also a Siamese junk which transported 14 horses to Siam.[42] The trading interests of King Narai ranged widely, and 'Siamese' ships, whether European, Chinese, or 'Moor' vessels, regularly sailed the waters of the Gulf of Siam, the South China Sea, and the Bay of Bengal.

In the meantime, King Narai was similarly involved in diplomatic contact with Susuhunan Amangkurat II of Mataram. In 1687 an envoy from the *susuhunan* to the court of Siam appeared on the east coast of Java. The VOC reported that the envoy had with him seven Javanese horses as presents for the King of Siam. The Javanese embassy to Siam may have been part of a diplomatic alliance against the VOC, which was becoming a more and more aggressive economic and military power in Southeast Asia. A little earlier in that decade, the Dutch had taken over Bantam (Banten) and relations between the VOC and Siam had been quite tense during the first part of the 1680s.[43] It is quite possible that, at this particular juncture, the Siamese and Javanese courts were using the horse trade as a cover for deeper diplomatic negotiations. In political and military terms, however, the Javano-Siamese diplomatic initiative did not seem to lead to anything concrete.

King Phetracha's Quest for Horses (1688–1703)

After King Narai's death many contacts with the west ended or took on a different shape and form. The new Siamese king was, at least in the beginning of his reign, hostile to the French, while relations between the court of Siam and the English East India Company had deteriorated badly by the late 1680s. French diplomats and traders ceased to come to Siam (Father Tachard's efforts to re-establish diplomatic relations between France and Siam notwithstanding), while the English East India Company also stayed away, the only British traders visiting Siam being some country traders based in India. The Dutch, however, stayed on and concluded a new treaty with the Siamese court at the end of 1688.[44] The trade of the VOC after 1688 was largely concerned with the export of goods such as tin and sapanwood, and the sale of Indian textiles to the Siamese court, but references to the Siamese court buying up horses from Java continue to be mentioned in the sources.

King Phetracha was knowledgeable about matters equestrian, having been Master or Equerry of the king's elephants before his accession to the throne. The Dutch even maintained that the king rode horses more often than he rode elephants.[45] In 1689–1690 the VOC granted passes to two vessels equipped by the king's men to buy horses in Java, at the request of the *phrakhlang* (Kosa Pan, the former Siamese ambassador to the court of Louis XIV). However, it appears that only one of these two vessels had returned to Siam by 1691. The king's horse-buyers of 1691 were advanced a sum of 20 catties 4 taels by the VOC to buy the required horses, and this debt was paid back in Ayutthaya by the king's treasurers.[46]

Why did King Phetracha want to obtain good quality horses? The clearest answer is expressed in a letter of 1696 from Chaophraya Kosathibodi (Pan) to the VOC Governor-General in which the Siamese minister explains that the court wanted horses for the king's 'cuirassiers' (armoured cavalry troops).[47] But the king appears to have wanted high quality horses for his personal use too (see below).

An earlier letter from the *phrakhlang* to Batavia gives another inkling of which kind of horse was wanted most by the Siamese court. In his letter of 1694, the minister specified that the horses must be of full size and height, be around three-four years old (or at the most seven-eight years old), have a well-proportioned body, and must be able to learn 'the Siamese ambling gait'. The *phrakhlang* complained in 1694 that thus far almost none of the horses bought from Java matched the court's exact requirements. The minister went on to specify almost case by case the reasons why certain of the Javanese horses already bought for the court were deficient. Of the thirty horses sent over to Siam by the king's horse-buyers recently, only one, a black-and-white stallion, seemed to accord with the court's specifications. Certainly the old Siamese horse manuals confirm that some of the horses suitable for a monarch had to be white, black, or black-and-white.[48]

The shipping of the King of Siam's horses involved not only the Dutch and the Siamese but a vessel sent in 1692 from Siam to Batavia to obtain horses was skippered by a Chinese. This may or may not have been the same Chinese, called 'Im', who was mentioned by the *phrakhlang* in 1695 as having been involved in King Phetracha's horse-buying activities in Java.[49] The horse-buyers themselves were, however, probably Siamese. At least their ranks and titles are revealed in some of the VOC documents, especially in the letters written originally in Siamese by the *phrakhlang* minister, and in the letters of the *opperhoofd* (VOC chief merchant).

In 1691–1692, the two horse-buyers had been Khun 'Wijtsjatwatti' and Mun In Sombat. A few years later, in 1699, a list is given of the whole party of 17 Siamese travelling on board a VOC vessel, comprising nine 'caloangers' (*khaluang* or courtiers), two servants, five stable boys, with a certain 'Ebrahim' as translator. The leading *khaluang* in the party were Okluang Yokkrabat,

Okkhun Thip Chula, who though of *okkhun* rank was the leader of the horse-buying expedition, Okkhun 'Peth-ijntra', Okkhun 'Ritsorasin', and Okkhun 'Peth Sang haen'. The other four were of *okmun* rank. These Siamese officials took with them 50 bahar of tin, 100 piculs of copper, and some Japanese camphor to sell in Java.[50] This suggests that on every trip commodities were taken to be sold in Java, with the remaining amount of money needed for the purchase of horses being advanced to the Siamese by the VOC.

King Phetracha sent another well-documented expedition to buy horses from Java in 1701–1702. The Siamese party, traveling on the VOC vessel *Pampus*, was led by Okluang 'Rha sambat' (Ratcha Sombat or Raksa Sombat?) and Okluang 'Pijtak tsijn nasaij', and further consisted of two officials of *khun* rank, two *mun*, and two *phan*. The Dutch document, a letter written by the *opperhoofd* Gideon Tant, identifies the two *okluang* and two *mun* as 'merchants', each with a servant. The two *khun* ('Choen Craij sinthop' and 'Choen Assawat tsaij'), assisted by the two *phan*, were horse-buyers proper. This shows that the Siamese court was concerned that there should be a clear division of duties. In Batavia, the merchants would sell the cargo of Japanese copper, tin, spelter (zinc or pewter), quicksilver, and elephants' teeth (ivory) which they had brought from Siam.[51] As for the horse-buyers, they presumably purchased the animals with a combination of money advanced to them by the Dutch and the money received from the sale of their cargo. [52]

King Phetracha appears to have been very keen to acquire good quality horses from whichever quarter he could find them. A VOC general letter of January 1697 reports that he made efforts to buy horses from the Coromandel Coast, China, Manila, and – with the Company's ships – from Batavia. The VOC therefore presented to the king Javanese horses as presents, and later gave him a gift of three Persian horses.[53] Persian horses were obviously highly prized and much appreciated by the Siamese court, but a large and steady supply was not to be had. In 1697, the Dutch *opperhoofd* Thomas van Son wrote to the Governor-General that King Phetracha had sent an Okluang 'Amanakhan' ('khan' denotes a Muslim, perhaps) and Mun Inthramat to the Coromandel Coast with a cargo of 22 elephants and 240 bahar of tin, to be exchanged for horses as well as textiles.[54] The horses wanted by the Siamese in India were most likely these 'Persian' horses, probably with Arabian blood.

In 1694 a Persian mare was sent to the King of Siam, because there was no Persian stallion to be had in Batavia. Later on the VOC was dissatisfied with the valuation by the Siamese court of the Persian horses given as presents to the king in 1696. The VOC gave the king three large Persian horses (along with some Javanese horses), but complained that, in return, the court gave presents which were of lesser value than the price of the animals and other gifts combined. The king wanted to give 612 guilders less than the true value of the horses, but at the same time the VOC wanted to get rid of the horses because they were expensive to feed and care for. The problem seems to have been that the Persian horses brought to Siam in 1696 were not good specimens, possibly because they were of the wrong colour, or were unable to dance and prance 'in the Siamese fashion'.[55]

Towards the very end of his reign, in 1702, King Phetracha sent forty men to Java, bringing presents to the *susuhunan* of Mataram, Amangkurat II, then also at the end of *his* reign. The gifts taken to Java included two elephants, twenty spears, copper cups, and Chinese inlaid mother-of-pearl boxes. The members of this expedition travelled on a barque (possibly Indian or Javanese, the skipper being a certain 'Sirij rajalela') and stopped first at Batavia. The king asked the VOC to lend any necessary help to his men, including a cash loan of not more than twenty Siamese catties (*chang*). It appears that this 40-strong Siamese diplomatic mission to Java was essentially a horse-buying expedition, but it brought gifts to the ruler of Mataram because the *susuhunan*'s help and agreement were necessary to the procurement of horses for the court. It is not known exactly what fate befell this expedition, but a VOC letter of August 1703 describes the return of eleven of the king's men on a VOC ship belonging to the Company and on a Batavia burgher's vessel. Ten of the king's men remained in Java. It is not known what happened to the other members of the 1702 embassy; perhaps they had already gone back to Siam on the ship which had taken them to Java in the first place. A Dutch letter of January 1704 reveals that the Governor-General refused to loan the Siamese in Java a sum of 1,000–1,500 *rijksdaalders* (rixdollars) but there were no immediate complaints from the Siamese court, presumably because King Phetracha died in 1703. [56]

Apart from desiring a large number of horses from Java, the Siamese court at the end of King Phetracha's reign also wanted Javanese women dancers to be brought to Siam from the court of Mataram. The Siamese asked the VOC to be of assistance in this enterprise, too, but the Governor-General and Council were not enthusiastic about helping, considering the Siamese court's demands to be 'burdensome'.[57] Again there is a dead end as far as the sources on this matter are concerned, because no further mention of the dancers seems to have been made in the Dutch documents. Interestingly enough, a French missionary document written by Gabriel Braud, also dating from the end of this reign, mentions that the old king liked to watch young girls dancing, and even danced with them. The missionary puts a sexual slant on the matter, saying that the king preferred to live a life of the senses with his young dancers, and by implication was neglecting the affairs of state, which were in the hands of his son the Prince of the Front Palace.[58] One may speculate that direct cultural ties between the Siamese and Javanese courts existed at this juncture, and that horses just as much as dancers were very much a part of this contact.

The Siamese horse-buyers went to the north coast or *pasisir* of Java (also known as the 'East Coast' of Java) to buy their merchandise. The 'beautifully-coloured black-and-white' stallion bought in 1694, along with the other 29 horses of that batch, were obtained with the help of knowledgeable Javanese from Jepara, Semarang, Cirebon, and Kalinjamat. In his letter to Batavia dated 1694, the *phrakhlang* wanted the VOC to arrange for men who knew how to buy horses to help procure them for the king's men in Java. No expense was to be spared once the desired kind of horse was found. The Siamese minister wrote that large troops of horses were apparently to be found farther inland in 'Dongdoet', on the way to Mataram, and in 'Castiassa' or 'old Mataram' (probably Kartasura). The Siamese court even sent painted pictures of horses to Java to serve as models but unfortunately none of these pictures have survived.[59]

A Steady Supply of Horses for King Thaisa (1709-1733)

The reign of King Thaisa was a period of intense commercial activity for the Siamese court, especially trade with China or through Chinese intermediaries. It coincided with a period when the private junk trade of the Chinese in Southeast Asia overall was in a flourishing state. For much of this reign even the *phrakhlang* minister was a man of Chinese origins.[60]

Not that the Siamese horse-buyers stopped going to Batavia and the 'East Coast' of Java during the short reign of King Sua (1703-1709), as there is evidence that Siamese horse-buyers were given an advance loan at Batavia in 1708. By 1709 they were again ready to send sixty horses to Siam. The VOC, however, was on bad terms with the court during this period. The Company withdrew temporarily from Siam in 1705-1706, owing to the lack of 'free trade' in the kingdom, particularly the Ligor (Nakhon Sithammarat) tin trade. The VOC was consistently disappointed in its Ligor tin trade, even though the Company had had a tin export monopoly there since the 1670s.[61]

The Dutch were always careful to keep on good terms with the Siamese authorities, including important officials or ministers, because the prompt issuance of trading documents, and indeed the very success of their trade in Ayutthaya, could depend on the goodwill of these officials. In 1710, Batavia sent two horses to the *phrakhlang* minister (presumably the Chinese *phrakhlang*), plus a heavy metal clock for use at his 'newly constructed pagoda or temple in Ayutthaya'. These two horses, however, were Persian and not Javanese. The new king (Thaisa) also received a gift of two Persian horses upon his accession.[62] Persian horses were again sent as gifts to King Thaisa, this time towards the end of his reign, in 1730. But the two 'Persian' horses sent to Ayutthaya from Batavia in 1730 were in fact Jaffna horses from Ceylon (Sri Lanka). During this period the VOC had stables and a stud farm on Ceylon, breeding horses of the 'Jaffna' strain which must have originated from the Persian horse. In a general letter of November 1730 it was stated that five stallions and five mares were sought for the Siamese king and the Japanese shogun.[63]

Native rulers, European East India Companies and Spanish Governors-General used fine horses as gifts or presents all through the seventeenth century and at least a part of the eighteenth century (see Chapter 6 of this volume). From a cursory look at the VOC Governor-General's general letters alone, there is a lot of data on horses as presents. For instance, a general letter of 8 January 1641 mentions that the ruler of Palembang had sent a Persian horse to the *susuhunan* of Mataram, hence the dispatch of a return embassy from Mataram to Palembang during that past year.[64] The Dutch themselves also gave horses as presents, not only to the Siamese kings of Ayutthaya but also to other Asian potentates. In 1671, the VOC gave a Siamese elephant and some Persian horses to the Nawab of Bengal as presents.[65]

Problems and complications dogged the King of Siam's horse-buyers. The tardiness of a vessel taking them to Java could mean a costly delay, as happened in 1731-1732 when the VOC ship *Berbices* could only sail slowly from Siam to Java. Alternatively, the horse-buyers often arrived too late to buy any horses, going home empty-handed. Sometimes the horses they bought in Java died on the journey to Siam, as happened in 1730-1731. The Javanese horses could apparently only be bought with the permission of the 'Javanese Emperor' (the *susuhunan* of Mataram). The VOC, for instance, was to make clear in Japan that horses, especially mares, were – on pain of death – not allowed to be shipped out of Java without express permission from the Javanese ruler.[66] The Javanese, then, valued their breed of horse enough to try to control the export of mares.

It seems that by the reigns of King Phetracha and King Thaisa the Siamese horse-buyers had become familiar with the geography of Java, and with the areas where horses could be bought. They also had become more demanding. The *phrakhlang*'s letter of 1694 to the Governor-General referred to above mentioned various places in Java where horses could be bought. Sources from 1716 reveal that Siamese horse-buyers had obtained horses from 'the royal residence Cartasoura [Kartasura]'.[67] A letter of 1719 from the Governor-General and Council to Wijbrand Blom (then *opperhoofd* in Siam) complains that the Siamese horse-buyers were a burden, and brought no advantages to the Company. The Siamese court protested to Batavia in writing, saying that the king's horse-buyers had been unable to buy any horses outside the northern coastal town of Semarang. The Governor-General maintained that the Siamese were at perfect liberty to do so, but in fact were given too little money with which to buy good horses. Therefore, they ended up with lower-quality horses. The horse-buyers would then mislead the court about the true circumstances, in turn causing the *phrakhlang* minister to make a complaint to the VOC.[68]

Conclusion

The total number of horses bought by the Siamese and sent on to Ayutthaya is difficult to calculate because data on horses *bought* on Java and horses *sent* to Siam appear separately in the various scattered documents. It is possible that not all horses bought were shipped to Ayutthaya. From a preliminary count, it appears that around 900 horses were sent to Siam from Java during the 1686–1735 period, and that at least 598 horses were bought and sent over to Ayutthaya for use by the court during 1709–1733, the period corresponding to King Thaisa's reign alone. For some years, Siamese horse-buyers went to Batavia and were advanced a loan of cash by the VOC. During this period, the sums advanced ranged from a minimum of 1,000 *rijksdaalders* in 1702, to a maximum figure of 2,150 *rijksdaalders* in 1730, or an average of about 1,760 *rijksdaalders* per loan.[69]

Records of the activities of Siamese horse-buyers in Java dwindle from the mid-1730s. This does not necessarily mean that the trade in Javanese horses stopped, but that more research needs to be done. In 1735 Siamese horse-buyers from the court of King Borommakot, who had travelled on the VOC vessel *Jacoba*, were allowed to proceed to the East Coast region of Java, stopping first at Batavia. This expedition was led by Khun Phichai Sinthop and Khun 'Thiep Pha Tjie'. That year the Siamese managed to purchase fifty horses, and were permitted to ship them home to Siam in two separate ships. The VOC allowed this in order to please the Siamese court, with which it had been at loggerheads over the lack of freedom of trade in Ayutthaya. The Siamese horse-buyers wanted to ship the horses back to Ayutthaya in separate lots, something to which the Dutch did not always agree. In December 1748, the Dutch Governor-General's general letter mentioned another Siamese horse-buying expedition going to Cirebon via Batavia.[70]

The arrangements to buy and transport Javanese horses worked out by the Siamese court and the VOC from the 1680s until at least the 1730s reflect the VOC's desire to keep the Siamese kings satisfied, in exchange for trade privileges and a steady supply of Siamese merchandise, some of which came to the Dutch via the royal warehouses. The Company was not always happy about the inconvenience of having to advance money to the king's horse-buyers and transporting the horses on VOC ships on a regular basis. The Dutch, however, did not refuse to cooperate with the Ayutthayan court, because they felt that their commercial interests in Siam would be served if good relations were maintained with the kings and their officials. Doing business in a court-centred environment entailed having to make sacrifices to please that court from time to time. The VOC came to understand this very well during its long stay in Ayutthaya, and the Dutch attitude is clearly reflected in their role as facilitators and intermediaries in the Siamese horse-buying expeditions to Java.

PART TWO

6 Colonising New Lands: Horses in the Philippines

Greg Bankoff

The horse is not native to the Philippines but its arrival there forms part of the large-scale equine migrations of the late fifteenth and early sixteenth centuries in which the animal colonised the Americas. Its introduction into the archipelago after 1565 both precedes and forms part of the Spanish conquest of the islands. The pedigree of the Philippine Horse has its origins in the South with the introduction of animals into the Sulu Islands brought from Sumatra, Borneo and Malacca during the mid fifteenth century.[1] However, the horse introduced to the northern and central parts of the Philippines has a much more dispersed ancestry. Here the animal's history forms part of the wider Spanish conquest and settlement of the archipelago. Semi-wild animals were obtained from Mexico and adjacent territories and brought across the Pacific. Known as *creoles* or *mustangs*, these animals were already the product of cross-breeding and hybridity being descendants, on the one side, from the Andalusian horse of Southern Spain whose origins can be traced back to Arab and Moorish mounts and, on the other, from Norse or Germanic stock with which they were inter-bred to give size and strength. In addition to the Sulu and Spanish horses, stock from China and Japan were subsequently introduced and bred on great stock farms established by the religious orders in the seventeenth and eighteenth centuries.[2] Numbers had evidently increased to such an extent by 1689 that William Dampier was able to report they were plentiful on Luzon and had gone feral on Mindanao. Half a century later, the horse was even mentioned in connection with overseas commerce.[3] The resultant inter-breeding over generations blended the animal's different pedigrees to create the native Philippine Horse as a distinctive breed. Horses, then, were as much colonisers of new lands as the Spaniards whom they accompanied and bore, and, like their riders, were forced to adapt to the new circumstances and environments they found themselves in.

Colonial Life

While the horse has long played a significant role in Hispanic society as an engine of war and a symbol of nobility, it came to occupy a somewhat different role in the imagination of the indigenous population.[4] Horse owner-ship, in fact, had become widespread by the nineteenth century with many peasant families reportedly having more than one mount. At the centre of the horse's daily existence was the human–animal relationship that largely determined the type and quantity of labour expected from it. There were four main types of activities by the turn of the twentieth century in which domesticated horses were employed: as draft, pack, saddle and racing animals. The small size and strength of the animal precluded its use for heavy haulage and in ploughing; work that was performed instead by carabaos (*Bubalus bubalis*) and bullocks. In particular, the horse's physique and kinetics rendered it unsuitable for employment in the mud of rice pad-dies.[5] As a means of transport and light haulage, however, it was considered 'indispensable': 'The horse constitutes here an absolute necessity, crafts, industry, Commerce, the entire population without distinction of class nor position, are obliged on account of the climate, to avail themselves of this form of transport even for the most insignificant of social and commercial relations.' In fact, it constituted 'the sole means of rapid [land] transport' at that time in the archipelago.[6]

Horses engaged in a variety of portage functions ranging from drawing the four-wheeled carriages and chaises used to convey officials and the well-to-do to pulling the *carromatas* (long, narrow two-wheeled vehicles often with a tilt), small horse-carts known as *calesas* or *tartanillas* and the carts (*carretones*) much in evidence about public markets and used extensively to carry groceries and like loads (Table 6.1).[7] However, their role in this respect was limited outside of urban areas due to the poor condition of the roads. Three highways existed on Luzon to link Manila with the provinces: the Northwest through Pampanga, Pangasinan to the Ilocos provinces; the Northeast through Nueva Ecija, Nueva Vizcaya to Aparri in Cagayan; and the Southern through Tayabas, the Camarines to Albay. Even during the dry season, the sad state of repairs rendered 'whole sections next to impass-able' and communications were virtually suspended during the wet because of the mud and lack of bridges. Often travellers were dependent on fragile bamboo rafts to cross swollen rivers, a hazardous venture at the best of times. As for byways, 'it need only be said that they are few and, as a rule, in wretched condition'.[8]

Alternatively, horses might be used as simple pack animals. In rural areas, the larger landowners used *carromatas* or carts to transport their produce to market during the dry season but poorer farmers used a dragged platform without wheels known as a *cangue* or *carrosa*. The transportation situation in the Southern Tagalog province of Laguna in 1887 is probably indicative of the conditions experienced in many rural areas. The lake from which the province derives its name is surrounded by a series of high mountains that restrict extensive agriculture to narrow riparian areas where the majority of towns are also located. Most people were dependent on the extraction of oil from the cultivation of coconuts. An average family produced between 10-20 *arrobas* (from 115 to 230 kilos) of oil a year that required transportation to market to sell.[9] Cartage, therefore, was a primary consideration and anything that hindered its realisation or increased its costs impacted seriously upon people's livelihood and especially their ability to pay tax. Horses were the chief means of transportation and landowners

Table 6.1 Number of Conveyances Registered in Various Provinces, Philippines, 1889.

Area	Carriages	Chaises	Carromatas	Carretones	Saddle/Pack
Manila City	637	419	726	263	790
Manila Province	24	29	573	87	998
Zambales	6	24	34	23	468
Tarlac	9		78	1890	1018
La Union	16	33	84	394	470
Camarines Norte	13	16	11	79	163
Cebu	80	428		1033	213
Iloilo	177	342	164	1429	1464
Capiz	41	19	12	22	297
Negros	103	112	182	329	737
Bohol	23	68		43	922
Totals	1129	1490	1864	5592	7540
Horses involved	2258	1490	1864	5592	7540

Sources: 'Estado del Numero de Carruages, Carros y Caballos, Varias Provincias: Bohol, Camarines Norte, Capiz, Cebu, Iloilo, Intramuros, La Union, Manila, Negros, Tarlac, Zambales' [The Number of Carriages, Carts and Horses, Various Provinces], PNA, Carruajes, Carros y Caballos, Bundle 7, 1889.

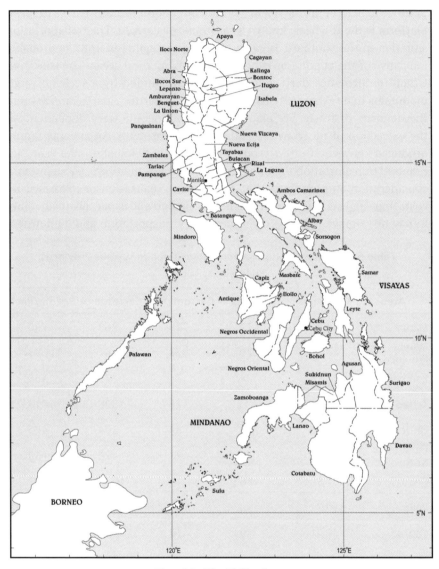

Map 6.1 The Philippines.

regularly kept several animals for this purpose. Even poorer families owned horses, regularly assisting their neighbours by providing animals without charge when there was need to transport quantities of oil to market. The animal's diminutive size, however, precluded it from bearing a load of more than four to six arrobas (46-69 kilos) on a single journey. If animals were not shared in this manner, then three or four trips were sometimes required to transport the 20 or so arrobas to markets that might lie as far as four leagues away. Apart from cartage, however, horses were seldom used for other purposes and were apparently let loose on the mountainsides for much of the rest of the time.[10]

In more mountainous regions, vehicles of all kinds were scarce and even saddle horses were used infrequently. A report on carriage ownership in Abra in 1889 identified only the provincial governor, the judge of first instance, the parish priest and the state-appointed doctor (*médico titular*) as possessing such a vehicle. Nor were there carts of any description as the precipitous roads of the interior did 'not permit the use of this form of locomotion in its main part' and there were no suitable roads that linked the province to the outside. Even saddle horses were rare being primarily used by officials on municipal business to the provincial capital. Few other people owned one. A similar state of affairs is recorded in Lepanto where there were no carriages or carts in the district and the few saddle horses all belonged to officials. But there may have been many more horses than such reports suggest. Considerable numbers of 'wild horses' certainly existed in highland Igorrot settlements in the interior of Luzon where they were permitted to 'run free in their fields without being able to determine their number'. Similar reports from Mindanao provide confirmatory data on both the absence of carriages, carts and saddle horses in Davao and of animals found in pastures cultivated by local *infieles* (literally pagans) who used them for travel and hunting.[11]

Of course, not all horses enjoyed such a seemingly unregulated or care-free existence. Increasingly the demands of a colonial economy rapidly being integrated into the world market created new types of requirements for energy that could often only be met through animal muscle. The proposal, for example, to establish a tramway between Campostela and a coal-mine located on the slopes of Mt. Licios on Cebu envisaged animals and not steam providing the locomotive power. Though ultimately mules rather than horses were recommended, such schemes nevertheless give some intimation of the new types of activities in which animals were increasingly employed. The plan called for the construction of a rail-line to transport coal and other materials in wooden trolleys with a capacity of about one ton at the rate of some 20 truckloads a day. Animals were required to haul this weight up 13 kilometres of mainly gentle slope over raised cement-like embankments built from calcareous stone and clay in the blazing tropical sun with little or no shade.[12] Similar fates awaited horses in the rapidly growing urban centres. In particular, the population of Manila more than doubled between 1814 and 1896 rising from 93,000 to over 190,000 inhabitants.[13] The territorial expansion of urban municipalities, their increase in population and the greater affluence of their inhabitants all gave rise to a demand for improved transport services. A tramcar system was first initi-ated in December 1883 with five lines radiating out from a central station on Plaza San Gabriel (now Plaza Cervantes) each pulled by a single horse. It was apparently not uncommon to see passengers disembarking to help the vehicles negotiate a slight rise or a sharp turn. And for the more well-to-do, there was always a sunset drive along the capital's *Calzada*, as much a social occasion as a scenic turn beside the bay, for which purposes carriages were hired on a monthly basis.[14]

A breakdown of statistics from various provinces gives some idea of the activities in which horses were engaged in 1889 (Table 6.2). While only indicative of the animal's employment, the data suggest that 20 per cent of horses were drawing carriages in comparison to 40 per cent used for haulage and the same per cent used as saddle or pack animals.[15] Many hire-carriage businesses were established, licensed to convey passengers around town, while others simply transported goods or baggage. Operating four- and two-seater carriages, these vehicles plied the streets within and the roads outside the municipalities from early in the morning to late at night.[16] A few of these businesses were quite large ventures such as the operation owned by Don Leocadio Roque on the Calle Real with seven carriages and 23 horses. Most, however, were more modest affairs involving 12 or fewer animals. Thus Doña Potenciana Villalon ran a hire-carriage business in Quiapo with 12 horses and Don Pulalio Acong had five carriages and five pairs of horses. Some were even quite small concerns, owner-operated or with one or two employees like that of Don Exequiel Toribio with his three pairs of horses and carriages and Don Lucio Santos with his two pairs of horses and carriages.[17]

Horses in the city were more likely to be housed in purpose-built stables. A description of one such structure, albeit built under the new US adminis-tration in 1905, gives some indication of the accommodation conditions of horses belonging to the growing class of carriage owners. The structure was 200 by 20 feet, provided shelter for up to 25 animals and used sawdust and dried grass as bedding material. But a particular feature of the equine colonial experience in the new urban areas, especially Manila, was the separ-

Table 6.2 Horses Employed by Activity, Philippines, 1889.

Carriage Horses	Haulage Horses	Saddle/Pack	Total
3748	7452	7540	18740

Sources: 'Estado del Numero de Carruages, Carros y Caballos, Varias Provincias', 1889; 'Supresión del Impuesto Sobre los Caballos, Abra 1889'; Blas Jerez to Director-General de Administración Civil, 1889; and Maximo Loilla to Director-General de Administración Civil, 1889.

ation of the sexes. A passing remark in a report to the Secretary of the Interior in 1903 makes the observation that 'nearly all the horses in Manila are males'.[18] Where statistical evidence of an earlier period is available, the data support such a contention. The province of Manila that can only be regarded as partly urbanised in 1889 recorded a population of 998 horses of which 864 were stallions (87 per cent) and only 134 were mares (13 per cent). Alternatively, the gender breakdown of the adult equine population in the rural municipality of San Pascual in Burias was much more even: 75 stallions and 67 mares.[19] As animal power was increasingly required to provide the sinews and muscle needed to support the economic expansion of the late colonial export economy, male horses were brought in from rural areas in the capacity of 'guest-workers'.

Racing was the other major activity apart from conveyance or haulage that involved the horse. Interest in the racetrack was keen with Spaniard and Filipino alike, the latter in particular observed to 'risk considerable sums on races of which they are very fond'. Some ethnic groups were also renowned for their horsemanship: 'that however steep may be the ascents they never alight'. Lieutenant Charles Wilkes, commander of the first U.S. scientific naval expedition between 1838 and 1842, observed how on Sulu, at least, riders used saddles cut of solid wood, preferring stirrups so short that they brought the knees very high.[20] By mid-century, the sport had achieved such a general degree of recognition among both colonised and coloniser that local race meetings were organised from at least the mid nineteenth century.[21] A Superior Decree of 12 June 1851 provided for public auction of the right to convene horse races in the province of Pampanga. Meets were to be held twice monthly on Thursdays so long as no cockfights were scheduled for that day. No formal racetracks as such existed yet but events took place at sites close to town to facilitate the maintenance of public order and alternated between the province's principal towns of Bacolor, San Fernando and Mexico.[22]

The formal origin of racing in the Philippines began with the establishment of the Manila Jockey Club in 1867, the first racing club in Southeast Asia. Initially races were run for purely recreational purposes without any betting. Races were held once yearly in April or May with the horses running on a straight track between San Sebastian and Quiapo churches, a distance of about a quarter mile. In 1880, the Club moved to a new site in a rice field abutting the Pasig River at Santa Mesa where meets of three consecutive days were allowed twice a year in February or early March. A permanent site was only established in 1900 when the Club leased and then later acquired 16 hectares at its present site at San Lazaro in Santa Cruz. The tremendous popular enthusiasm with which the introduction of betting in 1903 was greeted led to the establishment of a second racetrack in Pasay (Pasay Country Club). By 1906, there were no less than 220 racing days a year but insufficient public demand and the inordinate sums wagered on these occasions eventually led to the closure of the latter and to restrictions imposed on the former. Races were limited to the first Sunday of each month, legal holidays and the three days before Lent.[23] Native horses were much in demand at these events with some of the original 100 *socios fondadores* (founding members) of the Manila Jockey Club breeding stock for competition.[24] Winning times on the San Lazaro track between 1904–1910 show native horses achieving speeds of between 27 and nearly 31 miles per hour (43 to 50 kilometres per hour) over distances of up to two and a half miles (four kilometres). Larger mounts that showed promise of developing faster speeds consequently began to command prices of between P1000 to P2500 by 1911 or between 12 and 17 times their average market value.[25] More than anything else, racing and the status and money that accompanied it transformed the horse from a useful animal into a valuable commodity.

Colonial Health

The most significant event to affect the horse's health since its introduction in the fifteenth and sixteenth centuries was the unknown epizootia that emerged in 1887. Appearing first in the lowlands adjacent to the Pasig River in Manila and the Rio Grande in Pampanga, the infection rapidly spread far and wide during the *tag-ulan* or rainy season. Initially, this was a reference to the great bovine rinderpest epidemic that decimated up to 90 per cent of carabao herds during the late 1880s.[26] Yet there is a degree of confusion here as the reports clearly indicate that both cattle and horses were devastated alike. Symptoms, too, in light of later developments were ambiguous. Animals first appeared dejected, lost appetite, and exhibited signs of fatigue. Then their bellies swelled, they developed a cough, dry muzzle, diarrhoea and their respiration became agitated. In the final stages of the disease, the animals became emaciated, walked with difficulty and constantly lay down. Those that died of this disease had 'truly black blood, sticky, and refusing to coagulate and the intestines are dark reddish-black colour and marked with lacerations'. The cause of the disease was thought to be a parasite ('germ') that penetrated the stomach through the ingestion of grasses and plants infected by other animals' excreta.[27]

While horses are not vulnerable to rinderpest, they are to surra or trypanosomiasis, an infectious fever carried by biting flies (*Tabani striati*). An animal, whether healthy or infected, carries several hundred of these insects that remain with it more or less persistently for a distance of some kilometres. It is in this manner that the infection spreads. The disease causes emaciation, heightened respiration, high fever, paresis of the hindquarters that affects gait, leading ultimately to heart failure and death. Often animals simply fall to the ground and die of exhaustion. It affects not only horses, when it is always fatal, but cattle and carabao as well, who generally recover without treatment within three to four months. First noted in South America in 1842, the disease spread rapidly throughout the continent and beyond. It reached Mauritius during the Second South African or Boer War (1899-1902) and broke out in Java in 1900. It was considered to have first appeared in the Philippines in 1901 introduced either through infected cattle from the Netherlands East Indies or carried by race horses imported from India. The disease then spread

Table 6.3 Equine Population Changes, Philippines, 1886–1903.

Area	1886	1903	Change (%)
Luzon	203,012	108,067	−47
Northern Tagalog[1]	44,276	7,925	−82
Southern Tagalog[2]	60,813	49,259	−19
Pampanga	13,403	1,741	−87
Ilocos Region[3]	26,640	16,755	−37
Northeast[4]	27,468	10,005	−64
Bicol Region[5]	21,772	7,595	−65
Mindoro	7,015	2,505	−64
Visayas	24,423	26,782	+10
Cebu	4,634	8,427	+82
Panay[6]	8,667	3,403	−61
Negros	5,075	6,312	+24
Leyte & Samar	4,949	5,024	+2
Romblon & Bohol	1,098	3,616	+329
Mindanao	1,875	9,322	+497
Total	236,676	144,171	−39

[1] Bulacan, Pangasinan, Zambales, Bataan, Tarlac & Nueva Ecija
[2] Morong, Laguna, Infanta, Cavite, Batangas, Tayabas
[3] Ilocos Norte & Sur, Union, Benguet
[4] Nueva Vizcaya, Cagayan, Isabela & Saltan
[5] Camarines Norte & Sur, Albay, Sorsogon, Catanduanes
[6] Antique, Capiz & Iloilo

Source: Montero y Vidal, *El Archipiélago Filipino*, 1886, pp. 155–156, 329, 342, 361, 437; and *Census of the Philippine Islands, 1918*, Manila: Bureau of Printing, vol. 3, 1921, pp. 442, 734–735.

rapidly where its prevalence was reported 'from almost the entire group of islands' by 1903.[28]

However, it is possible that surra was already present in the archipelago prior to the twentieth century given the historical trade links between the Philippines and Spanish America, and that it was partly responsible for the earlier epidemic that affected horses in the 1880s. Such a claim was investigated and subsequently dismissed at the time as 'not of great importance one way or the other, except for its historic interest'. Whatever is the aetiology of surra and whether it had been present in the archipelago prior to 1900 or not, 'the frightful epidemic which has raged here is positively connected with this infection'.[29] The effect on the equine population was devastating: a comparison of horse numbers between 1886 and 1918 suggests the extent of mortality, especially among herds in the main centres of equine breeding on Luzon. The actual number of fatalities in some areas, however, was much greater than shown by the overall picture. A comparison of equine populations between 1886 and 1903 reveals the extent to which horses were affected in certain provinces and provides a tantalising snapshot of the extent of the epidemic as of 1903 (Table 6.3). Corroborating the account provided by the 1888 report that the epidemic had its origins in the areas adjacent to the Pasig and Rio Grande rivers, herds in Pampanga and the northern Tagalog region were reduced by over 80 per cent while those in the northeast, southern Luzon and on Mindoro lost over 60 per cent of their numbers.

In all, herds were initially reduced by nearly 50 per cent on Luzon and the disease may ultimately have been responsible for the loss of as much as 80 per cent of all animals by 1908.[30] The situation was less uniform in the Visayas where total horse numbers actually increased. The only exception was on the island of Panay where mortality levels reached rates comparative to those on Luzon suggesting that the epidemic had as yet only reached that far south. The statistics for Mindanao more probably reflect the greater reliability of the data since the effective incorporation of that island under the new American colonial administration. Still the figures are in line with the notion that the epidemic had not yet seriously affected equine demographic density there. It was not until 1918 that equine populations began to recover their pre-epidemic proportions (Figure 6.1). Even then, numbers did not exceed their 1886 densities except in the southern Tagalog provinces, the predominant breeding region in the archipelago. In fact, herds continued to decline in Kabikolan and on Mindoro. All these statistics, however, only record numbers for domesticated horses, those in a sense living within the bounds of the colonial administration. The really unknown factor is the size of the feral horse population.

Some intimation of the size of this other equine population is hinted at in later U.S. statistics on horses possessed by non-Christian tribes-people. There were 12,410 such mounts enumerated throughout the archipelago in 1903, a figure that had risen to 20,564 by 1918, an increase of over 65 per cent.[31] Even these animals should not be regarded as truly feral but more indicative of how horses had been incorporated into cultures that lay beyond the reach of the Spanish administration much as in the case of the Apache in New Galicia or the Mapuche in Chile. True estimates of equine populations lie largely beyond the horizon of historical sources and of historiography. However, the great surra epidemic was an occurrence of major significance in the historical evolution of the horse in the Philippines quite independently of its relationship to humans though the dramatic decline in numbers was a major, if still largely unrecognised, influence in shaping the course of the Philippine American War (1899–1902).[32]

Colonial Death

Like the human 'invaders', the horse also had to adjust to alterations in its environment, both physical and constructed. Natural hazards are features of daily existence in the archipelago: earthquakes, volcanic eruptions, typhoons, floods, droughts, landslides, tsunamis and the like occur regularly.[33] Floods, in particular, have historically been the source of much privation and suffering in the Philippines. While local accounts describe disasters that overtake human communities, they also occasionally refer to how other species fared during such events. Thus floods in 1839 and 1845 were described as 'so intense that many domestic and field animals were carried away by the waters of the river'. The typhoon that hit north-western Luzon on 25–27 September 1867 caused such a 'great flood' that more than 6,500 animals drowned in the rushing waters of the Abra River. The extensive flood of the Bicol River in October 1872 left the plain of Naga 'a sodden mass beneath a tideless sea,' where corpses, carcasses, carabaos, sewage and filth floated on the surface 'rotting in the sun, spreading pestilence in every direction'. One rare mention provides more detailed information on the flood that inundated large portions of central and northern Luzon in October 1871 and drowned 1,342 cattle, 761 carabao and numberless hogs as well as 842 horses in Ilocos Norte alone.[34] This carnage is equivalent to the total annual equine mortality for this area where statistics are available.[35]

Dangers, however, were also to be encountered in the rapidly expanding urban environment. The greater number of wheeled vehicles and the larger concentration of animals and people in confined and narrow thoroughfares inevitably led to increased collisions. Road accidents caused 22 human fatal-

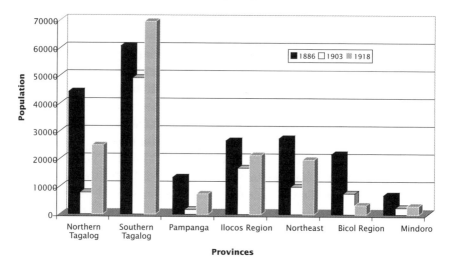

Figure 6.1 Variations in Equine Population on Luzon, 1886–1918.

ities in Manila during 1886, 21 of which occurred in the congested urban space of Intramuros. The walled city alone had 131 carriages, 56 chaises, 26 carromatas, 22 carts as well as 36 saddle mounts.[36] The number of animals involved in these crashes can only be surmised but horses were evidently sufficiently injured or maimed by such incidents to warrant their sale and early retirement from 'active life'.[37] Other animals or the conveyances they drew were abandoned or strayed, confronting municipal authorities with the problem of 'ownerless horses'. Thus police impounded an unattended cart on a public highway in Manila in May 1878 until surety and a fine were paid. Stray horses regularly wandered the byways especially in the suburbs and villages around Manila. On 5 June 1884, an animal was apprehended 'destroying the pasture' in fields belonging to Malacañang Palace, the seat of the Governor-General. Mainly, however, they were just found loose in the streets without anyone knowing to whom they belonged like the chestnut horse taken to the *Casa Tribunal* of Sampaloc. The usual procedure on these occasions was for an advertisement to be placed in the *Gaceta de Manila* describing the animal and calling on the owner to reclaim his or her steed. Those able to prove ownership were fined about two pesos before the animal was released to them. If unclaimed, horses were sold by public auction and often at bargain prices: Don Ramon Borras paid a mere P1.25 for the wayward animal that had browsed the gubernatorial pasturage. For those not so fortunate to be detained by the police, a worse fate could await. The *Guardia Civil Veterana*, the capital's police force, arrested three men late one night in April 1896 clandestinely dismembering a horse they had bought earlier that day in Manila without the requisite ownership papers. Men, meat and hide were later placed at the disposition of the *Capitan Municipal* of Parañaque.[38]

In more rural areas, theft was also a factor of life though carabaos were the preferred target more than horses. Thus Policarpio Rodillo, arrested by the *Guardia Civil* in 1891 and charged with the theft of 11 carabaos in Laguna, had previously been apprehended for stealing a horse.[39] Stolen animals were seldom recovered: beasts were peddled from door to door, taken to a neighbouring province, driven to market or slaughtered and disposed of locally as fresh and salted meat. Filipinos sometimes ate horse-flesh, usually dried (*tapa*).[40] The main onus of prevention inevitably fell on municipal authorities that were directed to rigorously enforce regulations regarding transfers of animal ownership as specified by the royal *cedula* of 16 November 1785. Buying horses or carabao without the knowledge and countersignature of the local *gobernadorcillo* (mayor) was strictly forbidden. Bills of sale required the vendor's name, date of sale and details of any identifying brands or characteristic markings. A copy of these bills was to be kept in a special town register and periodic reports sent to Manila. A late nineteenth century visitor, Alfred Marche, while observing how all farmers were required to carry a certificate of ownership for any accompanying livestock, wryly noted how 'thieves do not lack accomplices, which permit them to elude the order'.[41]

At the end of the road for horses lay deteriorating health, illness and ultimately death. The source of much common injury was incurred through service to their human masters. Sores to the top of the animal's neck, under the collar or breast strap, or on the back were mainly caused by harness and saddle and proved difficult to heal. Feet were particularly vulnerable to the hard surface of most country roads unless properly shod and maintained. Shoes were recommended to be replaced at least once a month as a growing hoof could throw the foot out of its proper axis and damage knee and fetlock. Even the 'very common practice' of washing animals removed natural oils from the skin leaving it scaly and the hair dry, dusty and rough.[42] Quite apart from the surra epidemic, other diseases also proved fatal. Glanders, a form of equine pneumonia, always resulted in death while anthrax, known colloquially as *garrotillo* or *carbunco sintomático* regularly took a heavy toll among horses within Manila. Animals passing through a big horse-market or transported any distance by ship usually came down with distemper, a type of influenza also known as strangles, shipping fever or stockyards fever. In mild cases, horses regained their vigour within a few days but sometimes complications might set in such as metastatic pneumonia or enteritis and result in death.[43]

Quite apart from these infectious diseases, there was also a host of non-communicable ones that constituted daily perils. Among many such conditions were heat-prostration or hyperpyrexia, common among animals in the capital especially during humid weather; laminitis or acute lameness frequently caused by hard road surfaces; fly-blow, the most troublesome of everyday conditions as wounds became infected by the larvae of a large blue-coloured fly (*Musca vomitorum*); and digestive disturbances or colic caused by indiscrete feeding or watering. Nor was poor health always the result of natural causes; animals were ill treated at times and strangulated, suffocated and submerged in water at others. A specific cost was even set for the examination of animals suspected of having been poisoned, with differential rates for those that demanded a more detailed chemical analysis and for those that involved only a simple autopsy.[44]

The health of horses had traditionally been entrusted to the care of the blacksmith. This person was typically a Filipino who combined the physical expertise of the farrier with the medical skills of local lore and belonged to that world of indigenous knowledge that lies largely below the horizon of modern historiography. The existence of this class of people is only occasionally hinted at in the records. 'In the Philippines, until now,' writes Felipe Jovantes justifying his continual preference for employing a local blacksmith to a professionally trained veterinary, 'it has been the custom of this country for there to be folk doctors (*medicillos*), unlicensed builders (*maestrillos de obras*), Chinese apothecaries, native herbalists and the like and no one interferes with them.' These native practitioners were skilful if the level of demand for their services is some measure of their worth, convenient in that they made house visits to their patients, and cheap, charging only modest fees in comparison to the professionally trained veterinarians. Among the treatments administered were bloodletting, the application of leeches and the use of acupuncture.[45] The 'professionalisation' of knowledge during the later nineteenth century, however, and the emergence of veterinary science as a specialisation replete with its own certified qualifications, specialist standards and fee structures had its counterpart in the colonial setting where native blacksmiths were progressively forbidden to practice this trade, prosecuted if they did, fined and even imprisoned. Care of the animal came a poor second to questions of inter-human competition despite the establishment of a health inspectorate known as the *policia sanitaria* charged with maintaining the welfare of livestock under the Royal Order of 13 April 1849.

Faced with the twin devastation wrought by surra among the equine population and the continuing rinderpest epidemic raging among carabao herds, the new US administration moved rapidly to establish a more comprehensive institutional and legal structure with which to tackle animal diseases. A veterinary division was inaugurated on 1 April 1904 consisting of nine veterinarians and 20 inoculators charged among other duties with inspecting animals entering the country. In its first year of operation, some 3,764 horses were inspected, 207 of which were subsequently condemned and cremated: 42 for surra and 165 for glanders. On 12 October 1907, the Philippine Commission enacted a quarantine law (Act No. 1760) making it an offence punishable by a fine of P1000 and/or six months imprisonment to knowingly import infected animals into the country or transfer them internally between locations. The law was bitterly opposed by cattle dealers who claimed its enforcement prevented buyers from acquiring the animals they needed for work or slaughter.[46] Permanent quarantine stations were located in Manila, Pandacan, Iloilo and Cebu. The creation of a College of Veterinary Science at the University of the Philippines in June 1908 and the provision of a board of examiners to supervise licenses completed the formal structure. An animal clinic was established in 1912 at a site between

Table 6.4 Causes of Mortality Among Equine Populations by Region, Philippines, 1918.

Area	Population	Deaths	Death rate (%)	Cause of Death (%)		
				Disease	Slaugh-tered	Other
Luzon	173,404	9,708	5.6	59	27	14
Manila	6,530	109	1.7	77	11	12
Northern Tagalog	24,917	1,905	7.6	49	39	12
Southern Tagalog	69,559	2,409	3.5	60	25	15
Pampanga	7,488	511	6.8	61	31	8
Ilocos Region	21,375	702	3.3	48	31	21
Northeast	19,762	2,075	10.5	59	26	15
Bicol Region	3,428	920	2.7	64	14	22
Mindoro	3,134	496	15.8	72	24	4
Visayas	39,296	3,485	8.9	52	42	6
Cebu	14,560	496	3.4	50	38	12
Panay	2,933	263	9	81	13	6
Negros	8,398	588	7	77	17	6
Leyte & Samar	7,191	1,890	26.3	41	56	3
Romblon & Bohol	6,214	248	4	49	38	13
Mindanao	14,036	1,814	12.9	77	13	10
Total	226,736	15,007	6.6	59	29	12

Source: *Census of the Philippine Islands 1918*, vol. 3, 1920, pp. 520–521.

Rizal Avenue and Calle Tayuman where consultation and treatment was dispensed free of charge. In its first seven months of operation, 457 animals were treated, the majority of whom were horses belonging to public-rig operators and the proprietors of livery stables.[47]

For the horse, the very end of a long road that could last 20 years or more came with death. The annual mortality rate for horses was between six and seven per cent of the entire population in 1918 though the regional variations were large (Table 6.4). In some provinces such as Leyte and Samar the percentage was nearly four times higher, while it was only about a quarter of that rate in places like Manila. Disease accounted for nearly 60 per cent of all deaths (8,880 fatalities), though again the rate fluctuated to over 80 per cent in some areas. However, not all deaths were the result of natural causes; nearly 30 per cent or 4,380 animals met their fates at the edge of the knife.[48] Old and unfit horses were often slaughtered and their flesh consumed. Licenses to maintain abattoirs in every town were sold by public auction and contractors specifically obliged to equip these places with good ventilation and adequate conduits for the removal of malodorous smells and the evacuation of waste liquids. Livestock was not permitted to be slaughtered anywhere else though owners were allowed to kill their own stock for home consumption so long as they paid the requisite tariff to the license-holder. Diseased animals were to be burnt along with their last place of confinement or buried at least one metre deep on beaches, sandy areas or other places not used for cultivation. On no account were their bodies to be thrown into rivers or estuaries.[49]

Colonial Legacy

Animals, however, are more than simply passive recipients of external stimuli and are also very much active participants in the process of historical construction, especially as agents of environmental transformation. It is difficult to assess to what extent the introduction of the horse affected the landscape, especially the flora of the Philippines. Human agency may have been the initial cause whereby the animal was brought to the archipelago but the degree to which it continued to be a factor of primary importance is hard to determine. Pastoralism, the human-directed management of breeding, was certainly a significant factor in localised areas and regional variations had already emerged by the nineteenth century. Horses from Batangas and Pangasinan were considered to be the best animals but those in the Bicol region were 'more delicate, although better adapted to racing'. They were 'small but strong' in Ilocos and rather wild in Jolo and Mindanao.[50] Though the horse spread to all parts of the archipelago, population densities varied considerably with enormous disparities between regions by 1886. Animal numbers were disproportionately concentrated on Luzon and the islands in its immediate vicinity (63 horses per 1,000 persons) in comparison to the Visayas (12 per 1,000) and Mindanao (10 per 1,000).[51] Southern central Luzon and the North became centres of horse breeding, most especially the province of Batangas, which alone accounted for between 11–14 per cent of all domesticated animals.[52] Other important horse-breeding areas included Cagayan, Ilocos Norte and Bulacan. As a general rule, there were fewer horses per person the greater the distance from Manila and/or the smaller the island. The only principal exception appears to have been the island of Mindoro with a horse population of 7,015 (1886) and a human one of 58,178 (1877 Census) that had a proportion of eight persons to every animal. In contrast, small islands such as Gunung Api off Sumbawa in the Netherlands East Indies (NEI) were commonly used for raising horses.[53] By 1886, there were an estimated 236,676 horses in the Philippines (an average one horse per 23 persons) compared to 1,079,699 carabaos (one per 5 persons), 555,016 beef cattle (one per 10 persons), 375,415 pigs (one per 15 persons), 15,547 goats (one per 375 persons) and 3,190 sheep (one per 1,740 persons).[54]

While reasonable statistics exist on the number of domesticated horses, the size of the feral population in the interior and its effect as an agent of environmental transformation is less clear. Bernice de Jong Boers notes the impact of large numbers of horses on the environment of Sumbawa in the Indonesian archipelago where equine densities reached six animals per square kilometre by 1930.[55] A working horse was thought to require over one kilo of food per day for each 50 kilos of weight.[56] About one half to two-thirds of this amount was given in grain and the balance in hay or grass. An adult horse, therefore, consumes as much as 60 kilos of grass a day, considerably affecting a landscape and contributing to deforestation.[57] The extent to which the introduction of the horse affected the landscape in the Philippines, especially localised ecosystems in those provinces of central Luzon where their populations were such as to make appreciable impact on the local fauna, has yet to be studied. They were, however, very much integrated into the agricultural cycle. In Catanduanes, for instance, horses were

Table 6.5 Domesticated Equine Population Densities per Square Kilometre, Philippines, 1886.

	Luzon	Visayas	Mindanao	Philippines
Horses per sq. km	2	0.60	0.03	1.16

Source: Montero y Vidal, *El Archipiélago Filipino*, 1886, pp. 155–156, 329, 342, 361, 437.

moved to upland communal pastures at the beginning of the rice-planting season where large areas of cogon grass were to be found, such as in the districts of Pandan, Viga and Baras. Once the harvest was in, horses were turned loose in the rice fields, now also regarded as communal pastures.[58] By the late nineteenth century, domesticated horse densities reached two animals per square kilometre for the whole of Luzon and over one for the Philippines as a whole (Table 6.5). Moreover, these figures do not include the unknown numbers of feral horses in the interior of islands.

Yet there are some indications of the possible changes that the animal wrought as an agent of environmental change. The absence of natural pasturage, the lack of extensive grasslands, necessitated that horses be fed on a diet that mainly consisted of a mixture of unhusked rice (*palay*) and maize though historically 'green provender' or *camelote*, a maize-like plant growing to a height of over two metres, was often substituted and, later still, oats when available under the US administration.[59] This nutritional regime, however, was considered insufficient to maintain improved bloodlines and even to adversely affect their progeny, who would lose the desired qualities of size and strength if raised only on 'the food regime of the country'. Instead, new 'artificial pastures' were required 'to procure more abundant and succulent food-stuffs than we possess now'.[60] As a first step, it was recommended that the seeds of more nutritious forage plants should be acclimatised so that the horses introduced from overseas could enjoy the 'same sustaining diet as previously' and so more readily 'counteract the effects consequent upon a change in climate'. In its 1883 submission, the special military commission even recommended the immediate introduction of barley, oats, alfalfa, sainfoin and piprigallo (both types of lucerne), carrots, and, above all, of *paraná* and guinea grasses which had given such excellent results in Cuba. It was also thought inappropriate to leave the success of this acclimatisation to the 'greater or lesser intelligence and interest of farmers'; rather cultivation of the plants should be entrusted to the expertise of agricultural engineers placed in charge of newly established state run model farms.[61] However, little advance was made in the creation of artificial pastures before the establishment of agricultural experimental stations under the U.S. administration in the early twentieth century.

Even less is known about the possible impact of the feral herds that Dampier witnessed or about the local practice of letting animals loose on the mountainsides when not required for light haulage or other purposes. The observation, however, that there 'comes a time each year when what passes as fields are exhausted and droves of hungry and gaunt looking mares go in search of food that they are unable to find' suggests a landscape, at least on a localised level, stripped of its undergrowth and denuded of foliage.[62]

In some parts of the Philippines, perhaps, animal and environment were bound in a similar relationship to that described by Melville for the Valle de Mesquital.[63] While the extent of this connection cannot be fully determined due to the lack of all but anecdotal quantification of feral numbers, the proper development of colts and the health of mothers were observed to suffer from such a degraded environment. In particular, colts, badly fed in their first year of life, were said to be enfeebled and especially susceptible to disease and mortality. Similar circumstances were to be found in other horse-breeding areas of Southeast Asia where the hot and humid conditions were blamed for low fertility rates and high mortality especially among foals.[64]

Like the Spaniards, then, the equine 'invaders' of the Philippines were the agents of considerable and long-lasting changes that substantially affected both the way of life of the archipelago's peoples as well as the landscape and ecosystem of the islands. In this, the history of the Philippine Horse like its counterpart in Java and Sumbawa is intimately linked to state formation and the colonial transfer of stock from within and without the geopolitical region. Yet, while it is necessary to appreciate the extent to which the animal was a significant agent of historical transformation in its own right and how its life was increasingly altered by its close association and integration into human societies, it is also important to realise that the horse itself did not remain unchanged by these processes but adapted over time to the different circumstances and environments it encountered in the archipelago.

7 Adapting to a New Environment: The Philippine Horse

Greg Bankoff

The horse was a newcomer to the Philippines, an interloper; more of an invader, in fact, than the Spanish who were principally responsible for its introduction and instrumental in its distribution.[1] After all, Spaniards were only a culturally distinctive type of humanoid and people had long been living in the islands. Horses, however, were a different matter, an altogether new species. Once established in the archipelago, though, sufficient inter-breeding over generations blended the animal's different pedigrees to create what came to be considered as the native Philippine Horse as a distinctive breed by at least the eighteenth century. In general, the horse's appearance is clean-limbed and rather upstanding with a broad head, straight face and small, fine ears, though those in the south were less affected by the sub-sequent importation of other breeds in comparison to those in the north and retained more of their distinguishing characteristics.[2] A particular feature of the breed is the manner in which the neck joins the body creating a comparatively straight line over the withers that tends to give the neck a longer appearance and causes the back to appear short. The latter is usually straight and narrow with a short croup and the tail set low. The feet are large, well-open at the heel, the knees are muscular, with well-laid shoulders though often the animal's hindquarters and withers are set back.[3] A particularly noticeable adaptation of the horse to its new environment, however, was its attenuation.

A Question of Size

Seventeenth century Spanish accounts compared the imported stock to 'draft-animals' and commended them for their 'well-shaped bodies, large fetlocks, large legs and front hoofs' and depicted the resultant crossbreeds with Chinese and Japanese stock as medium sized, strong, vigorous, and proven hard workers.[4] But as the number of horses increased, so the size of the animal apparently decreased.[5] Forty years later, an anonymous observer described the local breed as both 'numerous' and 'small', and its diminutiveness became the dominant characteristic noted by most subsequent Spanish and other European commentators.[6] As such, the native horse of the Philippines did not differ appreciably from other breeds around maritime or mainland Southeast Asia.[7] Similar descriptions are given of horses on the important breeding island of Sumbawa where the harshness of the local environment was both blamed for its diminutive stature and credited with its reputation for stamina and endurance.[8]

The increasingly diminutive size of the Philippine Horse was frequently commented upon by visitors and progressively regarded as a source of concern by colonial officials. 'The small size and inferior development of this country's horse', noted a report in 1881, 'is a cause of very great surprise to all who visit this territory for the first time.'[9] Cavalry Brigadier Sanchez Mira, while equally surprised, was more direct, calling the animal 'small and misshapen'.[10] While a few praised the animal's strength and endurance for its height, most were far less charitable, criticising its poorly developed chest, the shortness of its stride and its lack of stamina that necessitated draught and coach horses to 'require long and repeated rests to cover relatively short distances'.[11] Sanchez Mira was more censorious still: 'Equine livestock has degenerated to such an extreme in this Archipelago that it effectively now serves for nothing'. In particular, there were no horses capable of bearing heavy burdens (*tiro pesado*) and even very few saddle horses (*de silla*) and, as for riding the local horse, even 'if one is a horseman of regular height, it is impossible to mount one without looking ridiculous'. The question was

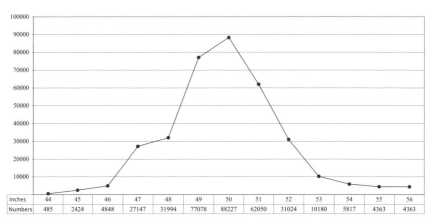

Figure 7.1 Comparison of Equine Height to Numbers, Philippines, 1935. Source: Nemesio Catalan, 'The Animal Problem of the Future Philippine Army', *Philippine Journal of Animal Industry*, vol. 2, no. 1, 1935, p. 70.

more serious with regard to the military. Local horses were not able to bear the weight of a rider and his equipment and it was estimated that only ten per cent of the cavalry would pass muster on inspection.[12]

Spanish measured horses in *cuartas*, the distance between a hand's outstretched thumb and little finger regularised as a length of 21 centimetres [8.3 inches]. By the late nineteenth century, an animal of between six and seven *cuartas* (1.26-1.47 metres or approximately 12.5 to 14.5 hands) was considered of 'great height' but there were few animals of this size and the majority were below six.[13] What this meant in terms of the horse population is suggested by a survey carried out in 1935 under the supervision of the US Animal Husbandry Department (Figure 7.1).[14] Measurements taken corroborate the earlier Spanish observations and confirm that the vast majority of the stock was small: 55.4 per cent under 50 inches (1.16 metres) and 96.6 per cent under 54 inches (1.25 metres).[15] An animal below this latter height was considered unsuitable for military purposes.[16] Certainly, the Spanish considered that the diminished size and strength of the horse was responsible for its limited usefulness, given the broken landscape, marshy nature of much of the lowlands and the total absence of roads in the greater part of the islands.[17] The subsequent American regime was to make the same disparaging remarks about the size of the local horse in 1900 as had been made a decade or more before, concluding that 'little care had been exercised in breeding them, and they might doubtless be greatly improved'.[18]

Human endeavour may have been responsible for introducing the horse to the islands but the resultant Philippine Horse was as much a 'mix' of different equine genes as it was a product of the environmental conditions it encountered. Attenuation was a very necessary physical adjustment to the sparse nutritional environment the animal found in the archipelago with its lack of natural pasturage for large animals. Reduction in size is not an uncommon occurrence for living forms faced with similar situations.[19] As a result of this loss of size, however, its utility for certain human activities also diminished. An important distinction, however, must be made between the needs of the colonisers and the colonised. For Spanish and American authorities, its diminished muscular abilities was a matter of considerable governmental anxiety but its relative size in relation to its usage by the indigenous population is more difficult to assess.[20] Regardless of this important distinction, however, the result was a determined effort on behalf of the state to reverse this natural process through policies that involved both the introduction of larger breeding stock from overseas and the wholesale emasculation of native stallions in an attempt to restrict their capacity to reproduce.[21] As the Philippine Horse adapted to its environment, it was itself subjected to change as a consequence of that adaptation.

Searching for New Blood 1881–1888

Size was to prove the most important factor in shaping colonial authorities' attitudes towards local horses. Official concern at the attenuation of the native breed had grown to such an extent that the governor-general convened a special subcommittee in 1883 composed of officers from the various branches of the military with vested interests in the animal to examine the reasons for its diminution and to recommend steps for its improvement.[22] 'The improvement of the breed of Horses in the Islands is no insignificant or trifling affair, nor is it a caprice just to waste money and put new burdens on the treasury; no, it is a justified necessity' wrote Brigadier Sanchez Mira once again, adding that it was 'indispensable' that the State woke up from its 'unproductive lethargy'.[23] There was general agreement that the animal's size had diminished over the centuries and it was the colonial government's attempts to reverse this process that gave rise to the decision to import foreign bloodstock from British India in the 1880s. The new US administration proposed similar measures to improve the native horse as early as 1902 and established a breeding station on Cullion Island, Palawan, stocked with American and Australian stallions and brood-mares.[24]

It had long been a widely held conviction in Spain that the native breed of horse there exhibited all the qualities most desired in the animal and so the necessity for cross-breeding or even for a trade in foreign horses (*carrera de caballos*) was 'stubbornly opposed' until the 1870s.[25] The Iberian Peninsula, in fact, remained largely isolated from the main currents of scientific thought until after the liberal Revolution of 1868 created an atmosphere more conducive to the dissemination and discussion of new ideas.[26] Under the reign of Alfonso XII (1857–1885), himself a keen promoter of agriculture and pastoral development, an equine breeding programme administered by the Ministry of War was established in 1869. Livestock owners sent their animals for servicing to the Agricultural Institute located in Andalusia. Many of the horses were of English pedigree, if names are to be taken as indicative of their country of origin, with stallions such as *National Guard, England-Glory* and *Norfolk Hero*, and breeding mares such as *Primrose, Big-Girl* and *Betty*. During 1882 alone, some 70 mares were presented at the Institute belonging to such notable breeders as the King, the Conde de Guagui, the Marqués de Castro Serna, the Duque de Sexto and the Marqués de Santa Marta. While the benefits of adopting this new practice were apparently so evident that there was scarcely a livestock owner of any importance who did 'not recognize the advantages of improving Spanish blood lines by crossing with English and Arab horses', the programme itself was regarded as having been largely 'ineffectual'.[27]

The situation in the Philippines was somewhat different, as the colonial state had already intervened on at least one previous occasion in an attempt to improve equine breeding. Between 1856 and 1858, there was an abortive attempt to improve the size of mounts available to the cavalry by establishing a breeding herd or *remonta* with imported Arab stallions.[28] However, failure to fully understand the dietary needs of horses and to provide adequate pasturage, together with a lack of proper veterinary support services resulted in most of the animals dying of starvation. Only two colts survived over the six years of the herd's existence and both died shortly after being pressed into service with the artillery battery.[29] The 'lamentable consequences' of this experience were held responsible for effectively curtailing any attempt to improve horse breeding for the next twenty years, leading many pundits to 'assure in prophetic tones that in these islands it is impossible to have good horses'.[30]

In its report to the governor-general, the special military subcommittee concluded that indeed the horse had reached such a state of degeneracy in the Philippines that it 'no longer satisfies any exigency, above all that of mounted agencies' and suggested that urgent measures be taken to improve the breed. In particular, the committee recommended the unusual step of importing both stallions and mares from Spain, the establishment of stud farms (*depósitos o paradas de sementales*) stocked with such animals, their replenishment every three to four years with fresh imports, the immediate purchase of Spanish mules for use by the mountain artillery batteries, and the creation of breeding herds if the number of mares should prove insufficient. Certain procedures were also advocated to avoid past mistakes, especially the need for acclimatisation centres to gradually accustom imported horses to local feed for a month before sale. Costs for the whole venture were not expected to exceed 300 pesos per horse and to be fully recoupable on sale.[31] On review, however, the Dirección General de Administración Civil disagreed with many of the subcommittee's recommendations. Citing the Ministry of War's unsuccessful breeding programme and the 'disgraceful complete and utter failure' of the previous attempt to import horses into the Philippines in 1858, it recommended instead that stallions and mares from British India be substituted for Spanish horses. Savings might be made from buying these animals at a lower price and by halving the costs and attendant risks of transportation. The state would restrict itself to acting as a purchasing agent on behalf of interested religious corporations and private individuals, recuperating costs on delivery in the Philippines. In the event, it was these proposals that were adopted and formed the basis of the Royal Order of 30 March 1885 granting approval for the project.[32]

Several years passed before any attempt was made to implement the programme and, by then, hope of substantial private participation in the venture had diminished. Nevertheless in 1888, a joint Commission comprising an agricultural specialist and a cavalry officer was dispatched to British India where it succeeded in purchasing 22 stallions and five mares.[33] Departing Bombay on 14 June, together with five ostlers and two months' provisions, the steamship transporting the horses was hit by a severe storm off the coast of Goa in which its shaft was broken and the craft left to drift until taken in tow by an English vessel to the port of Murmugão. A delay ensued to replace lost stores and settle salvage claims before the horses were re-embarked on another vessel bound for Singapore. News, however, that Hong Kong was gripped by a cholera epidemic induced the Commission to avoid that port and attempt the more hazardous direct route across the South China Sea to Manila where, after experiencing 'strong seas', they docked in August 1888.[34] Despite the vicissitudes of the voyage and sea-sickness almost always proving fatal to horses, amazingly only two animals were lost and the remaining 20 stallions and five mares were judged to have arrived in the Philippines in a perfect state of health.[35]

Evidently unaware of recent developments in the Philippines, the Ministry of Ultramar, the Spanish government department responsible for the colonies had already issued an order suspending the purchasing of horses in British India and recalling the Commission. Since it was not possible to comply with this directive, it was decided to sell the existing animals by public auction after a suitable period of acclimatisation.[36] Of the 25 horses put up for auction on 1 November, eight were sold on site, six more in the days following and the remaining animals sent to government agricultural stations, the Model Farm of Magalang in Pampanga (six horses) and the Agricultural Station at Iloilo (five horses).[37] The average price of 643 pesos paid per horse, based on the information provided for the six animals sold after the auction (four stallions and two mares) exceeded expectations and suggests that the venture might even have yielded a modest profit.[38]

If the nation state remained more an interested bystander in the scientific and medical advancements of the nineteenth century, content to leave such matters largely to the endeavours of private individuals and learned societies, its colonial counterpart was much more directly engaged in the practical application of new scientific knowledge, either directly or through the actions of its officials. The military and economic importance of the horse both as an engine of war and as a means of transportation ensured that the prevailing ideas on acclimatisation were of interest to colonial administrators. Faced with the manifest inadequacies of the horse in the Philippines, wrote Cavalry Brigadier Sanchez Mira, 'it is not possible to remain inactive, it is necessary to work, it is indispensable that the State take the initiative'.[39] Such views were only mirrored in the colonies of other European powers. Certainly Dutch, French and British colonial authorities were equally obsessed with a desire to improve local stock by breeding with imported animals and resorted to the same sort of expedients with largely similar negative outcomes.[40]

'Inventing' the Horse

As a consequence of expanding colonial empires, European natural scientists had become increasingly aware of the influence of terrain and climate on the distribution of species and had begun to speculate by the late eighteenth century on the historical effect that the exchange of flora and fauna between the continents might have had on the nature of those organisms. Both Buffon and Lamarck considered living things as inherently subject to limited transmutations in response to changes in the physical conditions of life and held that such acquired characteristics were inheritable.[41] The ability to manage this process of acclimatisation was recognised by the 1830s and the possibility of its profitable application had gained sufficient legitimacy for the state to promote its public utility.[42] In France, the centre of such studies, one of the most ardent proponents of the new science, Isidore Geoffroy Saint-Hilaire linked the process of progressive adjustment to what he called *zoötechnie* or the scientific propagation and breeding of plants and animals.[43] In an age of high colonialism, such considerations were to prove of some moment to European governments bent upon policies of territorial aggrandisement and resources exploitation between the Tropics of Cancer and Capricorn and to the animals that were to unwillingly or unwittingly accompany them. The attempt to improve the horse in the Philippines by importing animals from British India provides a means of understanding the way in which such ideas allowed Europeans to redefine the tropical environment around them during the nineteenth century not just as a physical space but as a conceptual one as well. David Arnold argues that scientific developments, especially the creation of medical specialisation in the diseases of 'warm climates', actually invented the notion of *tropicality* between 1890 and 1910.[44] The debate over equine zootechny, in particular the deliberations over what caused the diminution in the size of the horse, disclose the extent to which Spaniards (and later Americans) had begun to regard nature as yet one more tropical territory over which they could gain mastery and exercise dominion

The idea that humanity has a divinely-ordained right to dominion over nature is shown to be fundamental to Christian cultures.[45] The technological developments of the Industrial Revolution, with their emphasis on harnessing the forces of nature through the agency of the machine, did not seriously upset such a conceptualisation in Spain, let alone in her remaining colonies, where the ideas of Charles Darwin were largely ignored prior to 1868. Though evolutionary theories spread rapidly to 'even the remotest provinces of the nation' during the following decade, it was to Lamarckian notions of acclimatisation that colonial officials turned to provide the means of enhancing human mastery not only over inanimate but also over animate nature.[46] Animals, especially domesticated animals were represented as *locomotoras vivientes*, living machines or engines susceptible to improvement through breeding that was now 'no longer subject to obscure and uncertain laws but a science of calculation and growth'. Scientific breeding permitted humanity to obtain 'appropriate animals for particular services, with different aptitudes, according to what the caprice of the times demands or the necessities of Man'.[47] Some went even further, appropriating to the 'intelligence of man' almost the very act of creating beasts such as 'the horse, cow and pig, living machines that are aptly suited to obtaining the desired ends'.[48]

Science was to provide the means by which humanity could manipulate nature to compensate for any defect of function or pernicious effect of environment. The application of science to breeding was hailed as the answer needed to halt the continuing degeneracy of the native horse. Confidence in the new science was unbounded. 'Zootechny', wrote the special military commission, 'will give us the methods and sure rules so that improvements will not fail.' Nor was there any doubting humanity's right to benefit from such manipulation 'since the more it is possible to bring this living machine nearer perfection, the more will be the labours we can demand of it and the service with which it can recompense us'.[49]

There were three principal methods for improving the quality of a breed: selective breeding (*selección*), out-breeding in terms of bloodline (*cruzamiento*), and inbreeding or line-breeding (*progresión*). Selective breeding involved carefully choosing stud animals from among the existing stock for their desirable qualities and procreating from them. The method was considered practical only 'when the breed has not reached a very advanced state of degeneration' and was slow, requiring many generations to achieve satisfactory results. With out-breeding, mares from the existing stock were serviced by stallions drawn from areas of similar climatic conditions. The main advantage of this method was that changes in the breed were noticeable immediately and improvements were rapid. The last method, inbreeding or line-breeding simply comprised introducing stallions and mares from overseas stock and breeding exclusively from them. The method was quick but expensive and required the careful acclimatisation of the animals to be successful. As it was, the principal method decided upon in the Philippine case was out-breeding local mares with imported stallions: selective breeding was not considered feasible given the poor condition of the indigenous breed which would be unable to pass onto their progeny 'qualities which they do not have'; and line-breeding, though considered preferable in many ways and implemented on a very modest scale, was rejected because of 'its slowness and cost'.[50]

The proposal adopted fell far short of the comprehensive breeding scheme of state-stocked stud farms and an acclimatisation centre close to Manila initially envisaged by the special military subcommittee. However, the colonial administration hoped in this manner to meet the public demand for draught and saddle animals and to provide for the immediate needs of the military without over-committing itself economically. The decision to replace Arab stud stock from Spain with animals imported from British India was not purely a cost saving measure but also recognition of the importance of environmental factors. Horses from British India were already considered to be accustomed to tropical climate and pasturage and so were expected to more readily acclimatise to conditions in the Philippines. Much was made of the apparent successful adjustment of just such a stallion called (predictably) *Raja* and other animals already in the possession of a Manila resident, a certain Mr Barretts.[51] Nor would there be any loss in quality from such a substitution. The best bloodline in Spain, considered to be the *Casa de Monte* of the Carthusian Monastery at Jerez, shared many of the qualities of the horse in India, both being derived from the same Arab stock.[52]

The colonial authorities considered that these drastic measures were urgently required because of the state of degeneracy that the local horse had reached by the late nineteenth century. Its size and strength from their perspective fell far short of the ideal and was considered 'inappropriate for the service that this animal has been called upon to lend to humanity'.[53] But Spaniards were also aware that they had introduced the horse to the Philippines and that the present state and condition of the breed must be the result of local factors acting upon it. In particular, they identified the climate, poor breeding practices and the backwardness of agriculture. But if the animal had been changed once, then it could be changed back again, so to speak. And while the particulars of this process were to remain unclear until the mechanism of hereditary was properly understood, zootechny seemed to confer upon humanity at least a certain control over the forces of nature. Even as the tropics were increasingly perceived as hostile to those for whom it was not an ancestral environment during the nineteenth century, Spaniards progressively felt more able to 'mobilize the natural world to their economic and cultural advantage'.[54]

The tropical climate of the Philippines was initially held to be primarily responsible for the diminution in the horse's size. It was thought that constant exposure to the inclemency of the weather, especially the ultraviolet rays of the sun, had a 'modifying' influence on all animal organisms, changing their physical qualities and 'imprinting special and indelible characteristics that are transmitted hereditarily'. Ultimately, these processes could lead to the generation of a new breed, though the exact way by which this transformation was realised remained 'mysterious'. Despite the variety of climatic conditions prevalent on Luzon, some even similar to those in Spain, 'nevertheless in all these [areas], horses are small, badly formed and degenerate; evidence indeed, that makes it difficult to refute those who blame on the hot climate, the deterioration of the race'. Such views, however, were beginning to be questioned by the late nineteenth century: climatic conditions were still thought important but were no longer quite so generally considered to be primary determinants. Instead, the process of degeneration began to be primarily blamed on 'the pernicious system of breeding, caused by the lack of scientific knowledge, the want of abundant and adequate foods, and the backwardness of agriculture'. In fact, with advances in zootechnical knowledge and improvements in agriculture, the climate might even be converted 'into our ally, making it contribute to the generation of a breed of horses, that are naturalised here, and that pass on their good qualities to their descendants'.[55]

The change in attitude implicit in the development of these ideas represents an important conceptual shift from holding the tropical climate *per se* responsible for the causes of equine diminution to blaming current breeding practices and inadequate knowledge about nature for the present condition; in other words to meld the action (or lack of action) of the indigenous population with the environment as the agency of degeneration. Indigenous livestock owners were accused first by Spanish and later by American officials of following breeding methods best described as 'savage' and 'neglectful'. But this ignorance was not simply about maltreating animals, though Indios were accused of showing horses 'little affection' and not taking 'the care of them that such noble animals deserve'. Rather, the matter was one of poor consanguinity and repeated inbreeding, underlain by images of 'incestuous unions' and unbridled sexual licence.[56]

Obsession with the purity of equine bloodlines and prohibitions against inbreeding or crossbreeding have a long history in Spain where royal ordinances dating back to 'the time of the most ancient of monarchs' proscribed 'in truly horrendous terms' the propagation of mules ('that hybrid race') as prejudicial to the breeding of horses.[57] Such prejudices were still clearly discernible in the Philippines, whose livestock owners were accused of confining their animals together: mares of all different sizes, stallions without regard for condition or age, and their colts and fillies from the time of their birth to their sale in what was deemed by Spaniards as a 'detestable companionship' or more prosaically by Americans as 'co-herding'.[58] All were apparently kept together without any attempt at segregation 'without the intervention of the hand of man', allowing nature to indulge in a rampant permissiveness uninhibited by morality and governed only by the heat of the blood accentuated by the heat of the tropical climate. The result of this system of breeding was degeneracy, 'not a race that endures'.[59] Instead, what was evidently required was human control guided by scientific knowledge to eschew the effects of consanguinity and by the importation of 'superior' bloodlines as a means of improving 'degenerate' native breeds.

What all this meant, however, for the well-being and prosperity of the Philippine Horse was far more menacing. The threat to the native stallion was mainly to forestall its role as progenitor. While the body of the female horse was to be subordinated as the vessel for a new race, that of the male was to be assaulted by the wholesale castration of his kind.[60] Castration of native stallions not desired as sires, eventually to include all such horses, along with the introduction of new bloodlines from abroad, a process the Spanish referred to as *refrescamiento de sangre* (literally 'refreshing the blood') was deemed the most effective way to improve the breed, seemingly unaware of the irony that the same so-called 'Arab blood' that would improve the breed of horse had a long and bloody history of persecution among the human population of the Iberian Peninsula. Prizes were to be awarded for those specimens that most closely matched the desired conformational ideal. Nor were the Spanish alone in these beliefs: Dutch colonial authorities employed a surprisingly similar terminology to describe comparable practices. In the Netherlands East Indies, the process was referred to as *reinteelt* or 'clean-breeding' and its implementation involved the introduction of Arab and Australian studs, the selective castration of inferior stallions, the creation of a breeding register and the imposition of export restrictions.[61]

Castration was also regarded as having further 'benefits' from the human perspective. Far from weakening or making an animal sluggish, it was claimed by both Spaniards and Americans that a stallion became less distracted, saved energy that might otherwise be expended 'in fretting or other ways' and so enhanced his ability to work. In addition, even vicious animals became more docile and easier to handle but retained 'just as much strength and vigor as they had before'. In the case of young horses, castration was recommended at the age of one year.[62] The practice, however, was bitterly resisted by local Filipino horse owners: 'the prejudice against castration is so strong among the people that it is impossible to castrate more than a very small percentage of those that should not be serving mares'. Such resistance may have reflected practical concerns with ideas about virility and strength in both animal and human. While these policies remained largely notional under Spanish rule, the pressures brought to bear by the subsequent American administration proved much more difficult to resist. 'I have talked this matter over with some of them', reports the Director of Studs, H. Casey in 1905, 'and they are willing to try it.'[63] Even on the Catanduanes Islands, 'the work' was soon being carried out in all but one of the municipalities and even there it was 'shortly' anticipated to be introduced.[64] While castration is a recognised breeding technique and was vigorously pursued both in the metropole and by colonial administrations in many parts of the world, the planned wholesale emasculation of all native stallions was an attempt to genetically alter an entire breed of horse.[65] Colonial policies over equine reproduction represent a relentless intensification of human attempts to control the horse's corporeal form and reverse the animal's long-term adaptation to its new environment.

Horses, Race and Identity

The debate over the colonial government's attempt to improve equine bloodlines through a selected breeding programme with Arab stallions in the 1880s reveals much about changing Spanish attitudes towards nature in tropical regions. While colonialism had endured in the Philippines since 1565, it was only in the nineteenth century that Europeans began to see themselves as maladapted to settlement in the islands. The tropics were increasingly regarded as a hostile and deleterious environment, where prolonged exposure to a hot and moist climate was blamed for the poor health of individuals and for a progressive degeneration of the race. However, as increasing numbers of Europeans gained first-hand experience of living (and dying) in tropical regions, earlier notions about the gradual adaptability of physical constitutions to climatic differences through the selective adoption of local diet, dress and custom were replaced by the conviction that geo-medical boundaries restricted races to their ancestral environments. High white morbidity and mortality rates only lent credence to ideas about the unwholesome nature of the tropical climate and its degenerative effects on the physical and moral constitutions of the colonisers.[66] Such views were equally held with regard to the plants and animals introduced by Europeans whose relocation outside of their particular geographical region or *patrie d'origine* was blamed for progressive losses in fertility and diminution even when transported to similar climatic zones.[67] Nor could these beliefs really be separated from their moral overtones that seemingly endorsed polygenist attitudes about racial origins and provided a justification for imperial intervention.[68] And just as the tropical climate and indigenous population in the Philippines came to be held responsible for the plight of the horse, so, also, was the earth itself, the soil and what grew in it, found to be at fault, completing the portrayal of an environment inhospitable and malevolent to non-indigenous life-forms. Much was made of the backward state of agriculture: the poor nutritional value of available forage to sustain the proper development of colts and the health of mothers, and the complete absence of artificial pasturage.[69]

In short, then, the Spaniards held the tropical environment in its widest context – climate, soil, foliage and even native inhabitants – to blame for the diminution in size of the Philippine Horse. Moreover, it was an environment imbued with a moral laxity and sexual permissiveness, abandonment to whose forces could only lead to degeneration of both the physical and mental form. In both these respects, Spanish ideas differed little if at all from those of other European colonising powers. Views about the organic inferiority of the New World dated back to Buffon and eighteenth century America, while warmer climates had long been held to promote sexual license.[70] Nor did the matter stop with the beasts of the field. In the Philippines, the horse was considered as only suffering 'the consequence of the laws of nature that can be observed in man and animals of foreign extraction to these climes; they become debilitated, deteriorate in activity and strength and after several generations there is little utility to be hoped for from them'.[71] Thus the same environmental forces considered by the Spaniards to be at work on the horse were acting in similar ways on the human physiognomy and constitution. The tropics were defined as the region least suited to the white race: the ultraviolet rays of the sun were regarded as harmful to those whose racial constitution was 'potentially in harmony only with the particular ancestral environment, or climate, in which a race had evolved'; and the association of tropical heat with moisture and fever were thought to adversely affect human health.[72] Despite the enormous advances in medical knowledge consequent upon the discoveries of Louis Pasteur, Robert Koch and other micro-biologists after 1870 that diseases were caused by specific bacteria or protozoa, that is to say germs that conferred a certain amount of scientific respectability on such claims, prejudicial attitudes towards the tropical environment persisted.[73]

Spaniards also considered the constant hot and humid climate of the Philippines to be enervative and debilitating to the human constitution. Movement from the temperate climes of Europe to the moist, warm ones of the tropics was generally held to expose the European frame to an alarming range of ailments.[74] 'In a word, it is anti-vital to a greater or lesser degree, since that aforementioned temperature slackens animal fibre, producing a general laxity that with greater or lesser vehemence encourages lethargy and inaction with all their consequences.' But the effects were considered to go deeper than simply the health of an individual as the climate was thought to 'lessen animal life'. Diminutiveness in form was one of the results to be observed in the animal kingdom: 'weakness, decadence and smallness' diminished the higher one ascended the *ladder of life*, just as 'vigour, luxuriance and exuberance' multiplied the lower one descended towards the vegetable kingdom. That was why native 'ruminants, pachyderms and solipeds yield much in terms of size, weight, strength and gracefulness, to those from temperate zones'. But the same author held that 'equal or larger differences' existed with humans or what he referred to as 'bimanals', so that '[indigenous] people resemble so little the European with respect to their constitution, as much in their physical as in their intellectual capacity'. The results could even be seen among European residents of the tropics. 'But the difference does not end there; it continues yet among individuals of the same white race, that is to say, among the Europeans recently arrived in the country and those that have been two, three or more years residence. What profound and appreciable modification in temperament and level of energy the recently arrived cannot fail to observe [among the latter] in his first moments.' The only conclusion to be drawn was that: 'The Philippines is not salubrious for the European, who can be considered as a truly an exotic plant in that burning soil.'[75]

So the attenuation of the horse in the Philippines challenged cherished notions of Iberian identity, a culture that etymologically equated the status of gentleman with horse riding. The Spanish word for gentleman or sir, *caballero*, literally means horseman. To the Spaniard, the horse was a symbol of dignity and status: the King himself even issued detailed instructions about how an animal chosen to carry the royal seal re-establishing the high court in Manila in 1598 was to be caparisoned. The horse, a large gelding, was to have 'on the two sides of it hangings, which must be of brocade or silk, two shields bearing my royal arms, the face being covered with cloths of the same [material]'.[76] Above all, the furnishings of horses were a way of publicly displaying wealth and status in society. Accounts abound of the opulence of saddles decorated with ornaments of gold, precious gems, enamel and other rarities, and of gilded stirrups, rivets, bits and buckles. None more so than at the public festivities held in Manila to celebrate the accession of Philip IV, when the elite of Spanish society vied with one another in lavish displays of affluence. Few, however, could compete with the magnificence of Don Gerónimo de Silva bestride his great grey horse with its embroidered saddle and its band set with many pearls, the value of which was 'in the judgement of experienced persons, estimated at nine or ten thousand pesos', a truly fabulous sum for that time and place.[77] Spanish views linking horse ownership and status were still prevalent in the nineteenth century when Sinibaldo de Mas recommended the imposition of an annual tax on Indios and Mestizos who desired to use a carriage or saddle horse 'so that those who sustain this luxury may be very few'. Comparable attitudes have been noted of similarly martial societies who have made of the horse a potent symbol of military supremacy not only through making conquest possible but also 'through its own natural imperialism'.[78]

But the horse symbolised more than simply the dignity and status of the individual and at times came to represent the very majesty and puissance of the state itself. Governor-generals presented richly caparisoned horses to neighbouring potentates, as if the attributes of the animal were somehow representative of the majesty of Spain. One such horse, 'an excellent animal', was sent to the King of Cambodia in 1593 in exchange for the latter's princely gift of elephants.[79] The other circumstance in which the horse was closely associated with the power of the state was through the colonial army. The cavalry, in particular, was the most prestigious service within the armed forces. Indeed, a letter to the King from Governor Sebastian Hurtado de Corcuera in 1636 suggested that government expenses could be reduced by raising a company of fifty horsemen 'made up from the nobility of the city, who can keep horses'. Nor did the animal's association with the deployment of state power necessarily diminish with the passing of time and by the mid-nineteenth century the colonial army consisted of 1000 mounted personnel.[80]

Small horses that made an adult Spaniard look foolish undermined ideas about Iberian nobility and racial superiority. The question was even more serious with regard to the military. The attenuation of the horse over the centuries seemed to parallel the waning of colonial power and the animal's adjustment to its new environment to mirror Spain's need to come to terms with its diminishing position in world affairs. Such a decline had to be stopped at all costs according to Cavalry Brigadier Sanchez Mira. The attempt to improve the domestic horse through an aggressive breeding programme with Arab blood-stock imported from British India illustrates how Spaniards had begun to rethink their notions about the environment and their ability to control the forces of nature around them. Moreover, it clearly shows the important role the colonial state took in this enterprise. Far from substantiating the image of administrative stagnation that their colonial successors were only too eager to disseminate, Spanish officials displayed an easy conversance with the contemporary literature on zootechny and acclimatisation and exhibited a willingness to put it into practice.

Conclusion

Horses may appear at first sight a somewhat tenuous method of analysing broad changes in biological thinking during the period of high colonialism, yet animal breeding provides a means of understanding the close relationship between environment, science and imperialism during the nineteenth and early twentieth centuries. The widespread domestication of animals and their successful propagation around the world were frequently cited as examples in support of acclimatisation and the possibility of colonial settlement outside of original climatic homelands. Alfred Crosby depicts the impact of imported livestock on the 'Neo-Europes', the temperate regions of North/South America and Australasia as a 'world-altering avalanche'. However, the situation in Europe's tropical colonial possessions was somewhat different: rather than the animal transforming the environment, the environment transformed the animal; rather than proliferating in unprecedented numbers, its physiognomy progressively became more diminutive.[81] Yet Europeans often still remained dependent upon such imported livestock and the questions raised by their modification were of considerable military and economic significance.

The colonial state's attempt to increase the size of the domestic horse in the Philippines and other areas of South and Southeast Asia through outbreeding local mares with imported stallions and the concurrent debates over acclimatisation and zootechny, however, reveal another dimension to the period of high colonialism: the way in which some Europeans had subtly begun to redefine the natural world around them. Just as the tropics increasingly began to be perceived as inherently inimical to white settlement, confidence was growing in the colonists' ability to control those same forces of nature, enhancing their consciousness of superiority and right to dominion. To Spaniards, nature in the tropics was strange, having no clearly understood effects on animal physiognomies that were thought to retard growth, cause diminution in physical size and, by implication, mental development. Those to whom the tropics were not an ancestral environment degenerated, became enfeebled and debilitated and so served 'little utility'. And tropical nature had a still darker side, being perceived as malevolent, inimical to the health and welfare of the European who was 'truly an exotic plant in that burning soil'. Indeed, the idea that a cost had to be paid in terms of higher morbidity and mortality rates among Europeans stationed overseas had long been recognised and even monetarily quantified by colonial authorities.[82]

Of course, the way in which Spaniards perceived the tropical environment and its effect on the horse depended much upon their scientific understanding of the world about them. As Warwick Anderson has argued, technical knowledge of any kind constitutes a form of colonial appropriation with the ability to 'shape and regulate colonial social life'.[83] After more than 400 years of imperial venture in the New World, Spaniards were engaged in reinventing the Philippines by the late nineteenth century through a discourse on tropicality that redefined the islands and their own role there. No longer was their continuing imperium justifiable simply in terms of the 'one true religion' and the saving of souls; their new mission was cast in the language of scientific knowledge and as the purveyors of progress, a rhetoric that became more pronounced under the ensuing colonial regime. Spanish views on zootechny considered the local horse to have become degenerate, a condition blamed on the climate, poor breeding practices and the backwardness of agriculture. But if this state of affairs was the result of the tropical environment, these same factors were also at work on all animal life in the archipelago, including the human population. Indigenous peoples were first melded with the environment becoming indistinguishable from it, and then imbued with all the attributes associated with the natural forces around them: debilitation, dissoluteness, depravity, degeneracy and diminution. And just as nature needed the intervening hand of man, white man, to reverse the process of retardation, so the local population required the guidance and control of Spanish colonialism to escape from 'savagery'. In the process, the Spanish appropriated the natural world to condone and even justify racial superiority and European imperialism. And, the Philippine Horse? Once adapted to its new environment, it stubbornly resisted all attempts to increase its size and can still be seen today much as it was four centuries ago.

8 Riding High – Horses, Power and Settler Society in Southern Africa, c. 1654–1840

Sandra Swart

Just as in the case of Southeast Asia, the growth of the colonial state and the rise of the Dutch *Vereenigde Oost-Indisch Compagnie* (United East India Company or VOC) during the seventeenth century had considerable consequences for the human and equine populations under it, linking far-flung Southeast Asia and the southern tip of Africa. Although species of the genus *equus* – like the zebra and ass – have been present in Africa since earlier times, the horse (*Equus caballus*) is not indigenous, but was introduced into the continent. Although the horse was in regular use in North and West Africa from 600 AD, there were none in the southern tip prior to European colonisation.[1] Both African Horse Sickness and trypanosomiasis presented pathogenic barriers to horses reaching the Cape overland. Just as in the Philippines, *Equus* was an element of the 'portmanteau biota' that followed European settlement of southern Africa from the mid-seventeenth century.[2] The early modern colonial state that had come into being by the nineteenth century – against resistance from the metropole – was based, at least in part, on the power of the horse in the realm of agriculture, the military and communications. This chapter seeks to explore a particular facet of horse–human relationships, focusing on their introduction at the Cape and the consequent symbolic and practical ramifications. As in the case of Southeast Asia, the 'invention' of new horse breeds meant the dissemination of equine genes and phenotypes from Europe, Asia and the Americas and their fusion through deliberate state intervention, idealistic individual efforts by groups of breeders and often by the everyday politics of economic pragmatism. The resultant breeds were thus partly a product of their adaptation to new environments, and largely a corollary of their close connection to human society.

This chapter discusses the history of horses and white settlers at the Cape. Horses were the first domestic stock imported by the settlers and horses became integral to their identity as Europeans, used both symbolically and in a material sense to affirm white difference from the indigenous population.[3] The power relations played out through the horse are explored and contextualised with references to other settler societies. The socio-historical role of the equid in settler society and the divergence in horse culture from c. 1796 is discussed. The equine breeds became progressively more differentiated: the original 'South East Asia Ponies' made way for English Thoroughbreds imported for the racing industry, itself a consequence of British imperial interest in southern Africa from the late eighteenth century. The English thoroughbred followed a very different trajectory from the other key 'breed' of horse, which became known as the 'Cape horse', derived from original 'South East Asia Pony' stock with a globalised admixture of American and European stock.

As Necessary as Bread?

The Cape's strategic positioning, midway between Europe and the East Indies on contemporary maritime routes, and the region's agreeable climate rendered it potentially useful as a site for European settlement. In 1652 the VOC commissioned Jan van Riebeeck[4] to establish a re-provisioning station there to develop a meat supply and to cultivate fresh produce. In 1657, one year before the first import of slaves to the colony, the VOC released nine of its employees from their contracts, creating the first land-holding, free burgher community at the Cape.[5] They were intended to establish independent commercial farms that would provide the settlement with a steady food supply. The indigenous Khoisan[6] proved reluctant to enter the settler-controlled wage labour economy, which both exasperated and mystified the colonists.[7] This added to the perceived necessity for equine draught power.

There is evidence to suggest the existence of only one feral horse pre-Van Riebeeck, possibly the survivor of a shipwreck.[8] He was observed to be wearing a decaying rope halter, and was so wild he could not be caught. Consequently, the settlement initially faced construction with no draught animals. It was not part of the Dutch social tradition to use oxen for draught. Moreover, while oxen were occasionally ridden by some indigenous groups, controlled by a stick thrust through their nostrils, they were untrained in drawing wheeled vehicles.[9]

Van Riebeeck wished to reshape the landscape, to change the native ecosystem. The first white settlers attempted what is common to most settlers – to make themselves *at* home on the land by making it *like* 'home'. One aspect of this was the introduction of horses, which were not only an alien species from the metropole, but had the potential to transform the physical environment with their draught power.[10] Van Riebeeck wished to reshape the land itself with horses: he wanted to remove the bushes,

plough the soil, and cut down shrubs and trees, with that arboreal ani-
mosity common to settler societies, and encircle his settlement with a
hedge of wild almonds.[11] He argued that horses would prove invaluable
in transporting lumber, sand and clay and firewood to make bricks for
construction, and in revolutionising agriculture at the settlement with
ploughing and threshing.[12]

The Company, however, acceded to requests that delivered more immediate material results, and his written requests were ignored by his superiors
in Amsterdam.[13] Moreover, the original plan was not to create a colony, but
rather merely a refreshment station at which to refuel ships *en route* from
the Netherlands to the East. Horses and free burghers were inextricably
entangled in Van Riebeeck's changing vision for the development of the
country; with enough horses he believed he could provide provisions for the
Eastern settlements – and fashion a colony rather than merely a provision-
ing station.[14] He attempted to persuade the Company by arguing that only
horses would allow for the exploration of the hinterland. In May 1653, he
observed 'I wish we had a dozen [horses], when we could ride armed to
some distance in the interior, to see whether anything for the advantage of
the Company is to be found there …'.[15] He argued that it would accelerate
the settling period, saving both time and human labour (particularly after
he erected a horse mill in May 1657).[16] He further contended that horses
would make the settlers independent of the Khoisan, enabling them to
acquire their own construction materials and wood, thus freeing them from
an otherwise unavoidable policy of conciliation.[17]

Accordingly, Van Riebeeck argued that his biggest problem was the
continual shortage of horses. He wrote in his journal that horses were '*soo
nodigh als broot in den mont zyn*' ('as necessary as bread in our mouths').[18]
It was to become a perpetual refrain. He recorded in November 1654, '[I]t is
to be wished that we had a few more horses than the 2 we have at present,
both of which are being used for brick-making. We should then be able
to get from the forest everything we need, both timber and firewood, as
the roads are quite suitable for wagons.' In April 1655, he wrote, '[W]e are
therefore still urgently in need of another 6 or 8 horses.'[19] He judged horses
to be his 'greatest and principal need', demanding: 'Horses! Give me more
horses!'[20] His requests became petitions, which became pleas, but remained
ignored by Amsterdam.[21]

Zebra Crossing?

Van Riebeeck was hampered by two factors: the strict economy of the VOC
and the Company's desire to maintain the region as a refreshment station

and not to develop it as a colony in its own right.[22] He was thus refused, with the Company suggesting that he avail himself of the '*wilde paarden*', 'indigenous horses', the zebras and quaggas. Three wild members of the horse family were local to the area – later classified as the mountain zebra (*Equus zebra*), Burchell's zebra (*Equus burchelli*) and the quagga (*Equus quagga*) now extinct.[23] Van Riebeeck initially planned to tame them but found that he could not even catch them.[24] He recorded in 1660, that one of his explorers, Pieter Meerhoff shot a 'wild horse' [zebra], and while it was down straddled it in order to sever its sinews, but 'the horse rose with him still astride, and immediately jumped a stream ... and [Meerhoff] received a kick in the face.'[25] Van Riebeeck noted that the 'horse' was a 'beautiful dapple grey, except that across the crupper and buttocks and along the legs it was ... strangely streaked with white, sky-blue and a brownish-red. It had small ears just like a horse's, a fine head and slender legs like the best horse one could wish for.' He distinguished between quagga (which he observed the locals called *Douqua*), which he likened to mules, and zebra, which he likened to horses (and which the locals called *Haqua*). He tried to get living specimens of either but the Khoikhoi refused to assist, an indication of the dangerous nature of these equids, or, as Van Riebeeck argued, an indication that 'the [Khoikhoi] are beginning to realise more and more that we would thereby be the better able to keep them in submission'.[26]

The traveller Kolbe speculated, in 1727, that if domestic stock had not been introduced so rapidly, a more determined and sustained attempt at taming the indigenous 'horses' would have taken place.[27] Certainly, even after the introduction of horses and the founding of an embryo-horse industry, indigenous equines with their apparent immunity to horse sickness, proved an inviting proposition.[28] Anecdotal evidence has it that quagga were more tameable than zebra and only their extinction prevented their use as a viable alternative to horses. The Swedish explorer, Anders Sparrman, noted in 1785 that it was possible to tame zebras, and as they showed little fear they could be turned out with the horses at night to protect them against predators, and further, 'Had the colonists tamed them and used them instead of horses, in all probability they would have been in no danger of losing them, either by the wolves or the epidemic disorder [African Horse Sickness] to which the horses here are subject.'[29] He argued that horses were weaker in the Cape than in Europe, and quaggas or zebras would make better use of the dry pasture available.[30] A Cape Town guidebook noted resignedly in 1819, however, that 'the zebra is said to be wholly beyond the government of man'.[31]

Introducing the Horse

Establishing an initial settler equine stock was difficult. The long journey between the Netherlands and the Cape militated against sending Dutch horses, and the VOC resorted instead to sending stock from their base in Java, possibly from Sumbawa. 'Javanese' imports were small hardy creatures, thirteen and a half hands high.[32] They were also known as 'South East Asia Ponies', an amalgam of Arab and Mongolian breeds, their ancestors having been acquired purportedly from Arab traders in the East Indies.[33] Van Riebeeck criticised these first horses as too light, almost like English genets or insubstantial French horses (not sizeable, heavy-built and solid like the draught horses available in the Netherlands). Horses and white settlers were sent to the Cape in the same year 1652; the horses however were purportedly driven onto the island of St. Helena by a storm.[34] In 1653 four 'Javanese' (or more likely, Sumbawan) ponies were imported.[35] In 1659, sixteen more horses were permitted by the Company to be imported from the East in order '*om alle roverÿe der Hottentotten tegen te staan*' ('to put an end to theft by the [Khoikhoi]').[36]

The horses – together with a pack of hunting dogs – were imported, to a certain extent to inspire terror in the Khoisan, who were beginning to initiate raids upon the settlement.[37] Van Riebeeck argued that a watch of twenty riders would prove a sufficient deterrent.[38] On 7 June 1660, the settler authorities used horses to display settler ascendancy: '... the Commander, galloping along the near bank towards the farms of the ... Free Burghers, soon disappeared from their view. His purpose was also to demonstrate the speed of the horses, which caused great awe among them.' Van Riebeeck noted with satisfaction that the local population were astonished and impressed by horses, because of the 'miracles' of speed he performed with them.[39]

Van Riebeeck was drawing on an entrenched tradition, as horses were long associated in western Europe with the society of the elite and with the culture of hegemony. The horse had long distinguished the ruler from the ruled, with the rider a symbol of dominance.[40] The cost included not only the purchase price, but food, the transport of fodder and water, protection, structural shelter and space,[41] accoutrements (tack, training devices and tack-cleaning materials), shoes or hoof trimming, labour (grooming, exercise and training), and medical attention.[42] In Europe, the nobility's focus on a range of equine activities (mounted games, dressage, ladies' riding, hunting, carriage or coach-driving, or racetrack), and – as the western/non-western interface grew – an interest in exotic breeds like Arabians and Barbs,[43] led to a marked and ever-increasing differentiation in varieties of horses, bred for particular social niches. The complex, almost balletic, movements of what we would now call dressage, the style identified as *haute école*, swept Europe. Practitioners were largely of the aristocracy, who had the leisure and financial resources to pursue the *art* – rather than the utilitarian dimension – of riding.

A similar phenomenon was not observable in early settler society at the Cape and horses retained a utilitarian function until the beginning of the nineteenth century. Because most colonists came to South Africa in the service of the VOC, the majority of early settlers represented the lowest class of the Netherlands' and Germany's hierarchical societies.[44] Company employment was both hazardous and low paid, thus attracting the poorest elements of society to its service – men unfamiliar with *haute école*.[45] In a different way, however, horses did become a symbol of status within the evolving southern African communities. Horses gained a military use from 1670. There were – perhaps consequently – several attacks by Khoisan on horses; two horses were killed in July 1672, for example. It is not certain whether these attacks were motivated by the intent of obliterating horse stock, a desire for food or whether a horse acted as a proxy of white settlement, being perhaps one of its most visible and vulnerable manifestations.[46] The body of the horse was thus both a symbol of power and, perhaps because of it, a site of struggle. In the cultural context of animals and racism, Glen Elder, Jennifer Wolch and Jody Emel have argued that 'animals and their bodies appear to be one site of struggle over the protection of national identity and the production of cultural difference.'[47]

The first armed militia controlled by whites in southern Africa was established by the VOC. At first the Company had no large garrison and therefore relied on a few soldiers supplemented with local farmers and the indigenous people – who volunteered or were compelled to join a *kommando*.[48] The first *kommando* was established in 1670, which initiated the horse's military role – as opposed to simply draught and transport. By the early nineteenth century, the *kommando* system had become the dominant military mode.[49] The horse-based *kommando* system extended into politics, culture and social mythology.[50] Just as in the case of West Africa, the dominance of cavalry had more than solely military implications; it had an important influence on the character of political structure and social institutions.[51] Horse culture in South Africa reinforced the complex interconnection of military, economic and political power, coupled to social institutions. *Kommando* was part of the social machinery in the construction of settler, particularly Boer, masculine identity. However, despite desperate efforts, horses were not contained by white society; the equine industry did not remain in white hands. By the eighteenth century, the stud managers were often 'coloured boys' or 'Hottentot[s] or Cape boy[s]'.[52]

The Growth of the Horse Trade

Several factors militated against the successful introduction of horses to the Cape. Transportation was difficult – in 1673 a cargo of horses drowned, only two were saved, and in 1690 all horses died on the arduous sea voyage. On land, conditions were equally dangerous. African Horse Sickness (often dubbed 'distemper') was a perennial problem.[53] A febrile and infectious disease, it was the scourge of the wet season – an ailment believed by the settlers to be carried by evening miasmas.[54] There was no natural fodder – hay or grain – and forage was often of low quality. Predation by lions and other wildlife played a minor role, killing some of the prime stock and necessitating unmitigated vigilance. In June 1656, when his best stallion was eaten by lions, Van Riebeeck anguished: 'This has greatly inconvenienced us, when one considers all the work done by horses – one alone does more than ten men in pulling the plough, in carrying clay, stone, and timber from the forest ...' Horses occasionally died from settler misuse, and legislation was introduced to prohibit the premature use of colts and fillies, which underlines the importance of horses to the young settlement, and how seriously the administration took them.[55] As previously discussed, as the indigenous population realised the extent to which horses formed the white power base, there were attacks on horses. Arguably, stock theft became increasingly a factor as local communities realised the utilitarian value of the horse to them.

Despite these limiting factors, however, there was a steadily growing horse trade and, although overworked, these ponies bred successfully. By 1662, there was a herd of 40. They were now integral to the defence of the settlement, with 18 mounted men patrolling the border against cattle-raiding Khoisan. By 1667 there were 50 horsemen on watch duty. When one of these guards and his mount drowned while drunkenly fording a river, the mare was mourned more than the soldier.[56]

Horses ceased to be under the sole control of the authorities in 1665, when a public auction was held and free burghers were granted the right to breed horses. They continued to be a public asset, however, and were considered vital to the functioning of the settlement. The polity exercised a controlling hand. In May 1674, for example, a man was prosecuted for shooting his own (rogue) horse.[57] The rising stock saw an increase in quantity and a decrease in quality. Inbreeding began to affect the herds, certain birth defects – like weak hindquarters – were becoming obvious. In a 1686 *plakkaat* (decree), the governor, Simon van der Stel, tried to prevent the deterioration of stock, imposing a fine of 40 *rixdollars* if colts were used under the age of three years, and importing stud stallions in 1689 (allegedly from Persia).[58]

The number of horses continued to increase. In 1673, two more Javanese ponies were introduced, and in 1676 two horses and four mules from Europe. During the first 125 years of settlement horses were imported from Europe to the Cape only once, all the rest were from the East. In 1683 four horses were imported from Persia via Mauritius, in 1689 eleven horses from Persia. By 1681 the Free Burghers possessed 106 horses, and the Company 91 horses. In 1700, there were 928 horses. By 1715 the Company had 396 horses and the white settlers 2,325. In 1719, when the first crippling epidemic of horse sickness hit, the industry survived despite the loss of 1,700 horses. By 1744 the colonists had 5,749 horses. Hackneys were evidently imported in 1792.[59] Horse breeding continued at a steady rate, resulting in a 'breed' or, more accurately, a broad morphological type of horse that came to known as the 'Cape Horse'.

Unlike the horse in the American west,[60] horses did not represent freedom or wildness to the white settlers, but rather civilization.[61] This was because there were no indigenous feral horses to be 'broken in'. This meant that extremely difficult exercises in importation had to be undertaken and – even once there – equine existence in the colony remained precarious, threatened by disease and predators, both human and animal. Horses were not wild creatures to be tamed, but extensions of western civilisation to be nurtured and protected in order that they serve the white expansionist projects. The frontier consequently did not develop rodeo or related equestrian games – there were no contests as to who could ride the

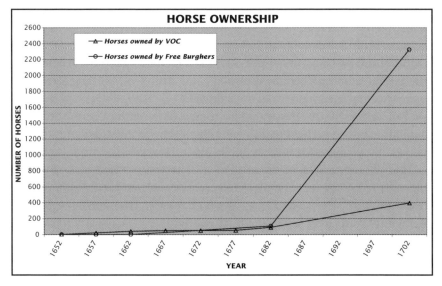

Figure 8.1 The Growth of the Horse Population in the Cape, 1652–1702.

wildest bronco. Horses meant (white) civilization – both symbolically and physically.

The 'Cape Horse'

It is a common trend that colonists create new breeds of horses to suit their needs. If a local horse existed (or if there were no indigenous horses but there were various breeds more readily available to import than those of the metropole), it could be utilised and often cross-bred with imported stock, and deliberately shaped, ultimately resulting in a new form of horse. These horses could differ markedly from those of the metropole, and after a while could come to be identified with the particular colonial culture, facilitating the differentiation from the metropolitan culture. After independence, horses were often one of the symbols utilised in the development of national pride and self-definition. This is illustrated vividly by the case of the gaited Peruvian Paso (a horse that walks and canters, but instead of trotting performs an amble, with legs moving in broken lateral pairs to create a four-beat smooth ride, like the *tripple*).[62] The Peruvian Paso, based on Barb and Iberian stock (including ambling horses) from the sixteenth century, were brought to the Americas by the Spanish. They were selectively bred for long comfortable strides, and were relied on for covering vast distances where wheeled vehicles could not go. From 1824, it became entangled in ideas surrounding national identity, and was therefore retained long after its functional utility had waned.

This general pattern was followed but with slight variation in the southern African context. There is little evidence to suggest an early identification of settler horse stock as different – and therefore superior – to metropolitan breeds. As we have seen, Van Riebeeck criticised these first horses as too light, and not solid like the draught horses available in the Netherlands. Correspondingly, there were ongoing efforts to alter the indigenising equine stock throughout the seventeenth and eighteenth centuries, with a view towards improving the horses of the Cape, who were becoming, over time, a phenotypical type or 'breed' called the 'Cape Horse'.

This 'Cape Horse' was a fusion of the southeast Asian or 'Javanese' pony (itself arguably of Arab-Persian stock),[63] imported Persians (1689), South American stock (1778),[64] North American stock (1782), English Thoroughbreds (1792) and Spanish Barbs (1807)[65] with a particularly significant Arabian genetic influence. In 1769, the first export of horses occurred; they were destined for Madras and initiated interest in breed-improvement (which encouraged the importation of new breeding stock).[66] Also in 1769 (and particularly after 1795, when enthusiasm for horse-racing proliferated after the British replaced VOC rule), the Cape-based horses diverged. One 'breed' to meet the needs of the racing fraternity, which used Thoroughbreds, usually a Thoroughbred sire and a colonial dam, and the other type to meet the needs of riding, transport and cavalry (which eventually became the 'Boerperd' or 'farmer's horse' or 'horse of the Boers'). This 'Boerperd' was variously known as the Cape pony or horse;[67] Caper (the name adopted in India for race-horses exported from the Cape); by English-speaking settlers as the 'Colonial', the 'South African', the 'Boer Horse'; and by Dutch-speakers as the Hamtam (an area in which horse breeding occurred); Melck or Kotze-horse (surnames of famous breeders); or even the *Bossiekoppe* (bushyheads or 'thickheaded'/unrefined) by those unimpressed by its points.

The 'Cape Horse' was phenotypically very different from English Thoroughbreds. They were small, hardy horses.[68] In 1845 the first 'scientific' description was offered: 14.3 hands, 'brown', compact, short-legged, calm, well-mannered, strong constitution, and disease resistant. For the first hundred years, the breeding stock were primarily Southeast Asian, with admixture from Arabian and 'Persian' stock, with an injection of English Thoroughbred blood from the late eighteenth and early nineteenth century. This globalised 'breed' was increasingly stamped as an indigenous horse, 'belonging' to the Cape.

The traveller George Thompson noted, for example, 'I had travelled this day about 56 miles, the last 30 at full gallop on a hardy African pony, saddled for me fresh from the pasture. This would have killed almost any English horse, but the country breed of Cape horses is far more hardy than ours ...'.[69] Lady Anne Barnard recorded her favourable impression of the Cape horse in 1798: 'A little more size in the breed would render the Cape horse very good; they already have a cross of the Arabian fire and are hardy in the greatest degree ...'. Daniel van Reenen was breeding Cape horses near the Breede River. Heinrich Lichtenstein noted that these horses were 'so fine, that they [were] very much sought for at the Cape as riding-horses' and '[t]he Africans, besides, owing to their being accustomed from their youth to seek their nourishment upon dry mountains are easily satisfied, and grow so hard in the hoofs.'[70]

Moreover, settler riding styles replaced those of the metropole.[71] This developed during the eighteenth century, which has been argued to be the period in which, with the decrease in settler immigration and increase in the births of Dutch-speakers, the Cape truly ceased to be an extension of Europe and became instead a *colonial* society. The trot, for example, considered 'unnatural', was replaced by the ambling jog discussed earlier called a *triple* (or *trippel* or *tripple*), or slow gallop, making it easy to ride while carrying a whip or gun.[72] Lichtenstein commented on the 'short gallop' of the Cape horses, 'very agreeable to the rider as well as to the horse ... This pace appears so natural to the race of horses in question that it is not without some difficulty the riders can ever get them into a trot or walk.'[73] In 1861, Lady Duff Gordon termed it a 'shuffling easy canter'.[74] H. Rider Haggard makes mention that 'he put the tired nag into a sort of "tripple", or ambling canter much affected by South African horses',[75] and that the horse 'reduced its pace proportionately, to a slow tripple, [...] Anscombe kicked the horse with his sound heel and I thumped it with my fist, thereby persuading it to a hand gallop [canter/slow gallop]'.[76]

Yet, over time, the Cape Horse was seen to deteriorate. From as early as 1800, there was growing concern about improving the Cape Horse. For example, the English official, William Duckitt, who arrived to establish an Agricultural Department and an experimental farm at Klapmuts, in September 1800, started a small stud of mares, and went 'native', becoming increasingly absorbed into the Dutch-speaking sector and concerning himself with improving the quality of the Cape Horse.[77] He wished to improve not the English Thoroughbred, but rather the draught horse. His argument was that the horses were so poor that farmers were reduced to utilising oxen for farm work. This meant that oxen were sold for beef only when past labour, and were therefore very tough and unpalatable.[78]

From 1860, breeding with inferior imported English Thoroughbred stallions was widespread enough to impact negatively on the export trade. Popular demand encouraged breeding from English Thoroughbreds, and pedigree became paramount to the exclusion of merit.[79] Speculators imported broken-down Thoroughbreds with good pedigrees.[80] These 'blood weeds', as they became known, polluted the gene pool, precipitating a general decline in stamina, size, hardiness, and sound conformation.[81] Moreover, particular horses were bred from because of the vagaries of fashion – white socks, a blaze on the face and chestnut coats were all unpopular, for example. Efforts towards breed improvement were further hindered by a focus solely on the sire to the exclusion of the dam. Interestingly (as a reflection on the gendered use of horses), particularly as good horsemanship was perceived to be a hallmark of masculinity, stallions received special treatment. Evidence is sketchy for the eighteenth century, but by the nineteenth century, it was considered 'undignified to ride or drive [mares]. Only Hottentots did it.'[82] The traveller Charles Thunberg noted in the late eighteenth century that while mares and colts were expected to forage for themselves, stallions were fed specially cultivated barley. Moreover, mares were perceived as having little importance and, as it was considered undignified to ride one, fillies were seldom broken or handled.[83] Similarly, fillies were rarely deemed worth racing. Wyndham notes similarly that in efforts towards breed improvement the infusion of 'good' blood was from the stallion, as 'any female equine was considered good enough to serve as a brood mare, and comparatively few were imported during the first half of the nineteenth century'.[84]

There was apparently little anthropomorphism surrounding the horse as 'personality'. In one particularly revealing incident in 1773, an official in the service of the VOC, Wolraad Woltemade (1700–1773), endeavoured to rescue sailors going down with a storm-damaged VOC ship. Woltemade made seven trips to the ship, riding his horse 300 metres into the sea, saving fourteen sailors. On his last attempt, the remaining sailors panicked, and too many clutched the horse, dragging them all down. Significantly, the horse was not accorded hero's status (indeed initially neither was Woltemade),[85] and the horse's name is lost to posterity. Correspondingly there appear to be few contemporaneous equine luminaries globally.[86] However, by the time of Dick King's famous 1842 ride from Durban to Grahamstown (a ride of ten days and 600 miles), the name of his horse, Somerset, grew to be well known to the public.[87] And by the late nineteenth century, certainly, there were various southern African horses accorded celebrity status.[88] It was arguably only with the advent of the racing industry from 1796 that horses began to acquire individual public *personae* in southern Africa, with race favourites becoming known and adored by the crowds. This was further stimulated by the rise of popular media, available more cheaply to an increasingly – albeit slowly – literate public; and further fostered by military campaigns (for example, 1881 and 1899–1902), which facilitated the popularising of heroes – both human and equine. This was further stimulated by the global trend toward popularising military heroes' horses, particularly after Napoleon's 'Marengo'. A similar argument may be made for England, and, gradually, the global context.[89] This became widespread with the 1877 publication of Anna Sewell's *Black Beauty*, serving as indicator and further promoter of the anthropomorphised horse as personality.

Race (and Class) Horses

By the late seventeenth century, the English had developed a keen interest in horseracing, a dramatic divergence from the *haute école* still embraced on the continent. This new pursuit required a new seat (or position in the saddle) and ultimately a changed equine physiognomy. The Thoroughbred came into being, adapted both conformationally and in terms of disposition for galloping. The characteristic for which they have been selectively bred is simply speed. For this they need to be leaner, of lighter build and longer-legged (almost a retention of neotenous traits) – and more vulnerable and less suitable for farm or draught work – than other horses. Thoroughbred breeders drew neither on the horses of the *haute école* nor on those of draught stock, but on the blood of the Barb, the Turk and the Arabian.[90] As Wyndham noted, '[H]orse-racing will start sooner or later in any country occupied by the British.'[91] The first English Thoroughbreds were imported in 1792, immediately prior to both the British Occupation and the institution of racing.[92] Although the Cape was not yet part of the British imperial web in 1792, an increasing interest was being taken in the English Thoroughbred, which was gaining international fame. In 1795, the year in which the British began their first occupation of the Cape, an Agent for the British East India Company imported more blood stock, which was followed by importation by the wealthy Dutch-speaking sector, particularly the horse breeders Jacobus, Sebastiaan and Dirk van Reenen.[93] A racing club, the African Turf Club, was organised and the first race meeting was held in Cape Town in 1797.[94]

There were no bookmakers and each bet was an arrangement between two independent parties. Initially, with the dearth of Thoroughbreds specifically for racing and races being held only once every six months, any horse could be utilised as a race horse: a hack, a hunter, a trap-horse. There were few horses, so the race-going crowd was able to identify the individual horses and, as mentioned above, had a greater personal knowledge of – and financial stake in – the horses. The races became a diverse space, mixed in terms of gender, class, ethnicity and race. The biannual race meetings were significant social events, incorporating assemblies and theatre for the elite, and drawing a varied crowd at the meeting itself. Initially, meetings were held on open commonage, with no fence and consequently no gate money. Burchell commented in 1810 that it was 'amazing' to watch

> the motley crowd on foot: Malays and negroes mingled with whites, all crowding and elbowing, eager to get a sight of the momentous contest. But the patient Hottentot … seems to prefer a pipe of tobacco to that which affords such exquisite gratification to his superiors. Together with the art of making horses run fast, the science and mystery of betting has found its way to the farthest extremity of Africa; and on Green Point large sums are said to have been won and lost.[95]

Lady Duff Gordon records:

> [At the Greenpoint] races, … a queer-looking little Cape farmer's horse, ridden by a Hottentot, beat the English crack racer, ridden by a first-rate English jockey, in an unaccountable way, twice over. The Malays are passionately fond of horse-racing, and the crowd was fully half Malay: there were dozens of carts crowded with the bright-eyed women, in

petticoats of every most brilliant colour, white muslin jackets, and gold
daggers in their great coils of shining black hair.[96]

There was a clash of horse cultures. Young subalterns, drawing on trends
from the metropole, docked horses' tails, which infuriated Dutch locals,
who observed that the tails provided protection against flies.[97] There was
no simple division between English- and Dutch-speakers over racing. More
complicatedly, there was a seam of English-speaking colonial society that
distanced itself from racing, principally because of its ties to gambling.
Governor Lord Macartney, for example, despised racing. Lady Anne Barnard
attended the first race meeting in 1797, while her husband absented himself
on the spurious grounds of official business. She was aware of the ambigu-
ities of her position – she personally disapproved, as did the governor, but
she attended in the carriage of a Dutch-speaker to emphasise her connection
to the broader population, and to show a lack of snobbery.[98] However, there
is no evidence to suggest that Sir George Young, governor from December
1799, disapproved of racing. Indeed he joined the African Turf Club.

Racing initially depended heavily on the garrison (the Turf Club began
with 29 members, 20 of whom were officers in the Army or Navy, with
only a few wealthy Dutch-speaking horse breeders). There were efforts
to encourage racing enthusiasm among the Dutch sector, by organising a
'Farmers' Race'.[99] During the period of Batavian Republic rule at the Cape,
1803–1806[100], there is little evidence of racing and the African Turf Club
went into liquidation. After the second British Occupation in 1806, Dutch-
speakers participated increasingly actively as spectators at meetings. After
1807, although racing depended heavily on officers, the farming sector
(particularly Duckitt, Melck, and the Van Reenen brothers) was progressively
more involved. The Turf Club provided the social space for the breeders
to meet and trade. Racing became increasingly popular, with race clubs
starting in Paarl (1815); Uitenhage and Stellenbosch (1816); Graaff-Reinet
(1821); Grahamstown (1823); Somerset East and Swellendam (1825). The
colonial governor, Lord Charles Somerset, was even able to counter imperial
rumblings over banning racing – because of concerns over gambling – with
the assertion that this would alienate the Dutch sector of the population.[101]
In 1822, for example, the Burgher Senate commented with approval on race
meetings and granted money towards a Cup, valued at 400 *rixdollars*.[102]

Class division was materially evident in the display of carriages at the race
meetings. Cowper Rose observed that one could see 'the regular gradation
from the well-appointed English carriage to that curious piece of antiquity
[Dutch cart] – the gig, the light wagon cart, and the long heavy wagon …
hired for the day'.[103] In 1820, William Bird noted that settlers, who could not

afford a horse in England, got a boost in status as they could afford one in the Cape.[104] Burchell observed that social display formed an important part of the events, 'Horsemen, without number ... exhibit their prancing steeds of half Arab or English blood; although some, indeed, of their noble animals refuse to prance without the incitement of the curb or spur.'[105] Similarly, class differentiation was marked with the vehicles acting as nuanced social signifiers, as indicated by this comment in 1822: 'lawyers in barouches; the next in rank in curricles, and notaries [following] in a solitary gig. Doctors turned out in chariots drawn by four greys, surgeons in barouches or tilburies [delicate two-wheeled vehicles], and an apothecary on a hack'.[106] Fashion vacillated, as Wyndham observed, with fewer carriages at meetings in the 1840s, but regaining popular appeal as a space for display by the 1880s.[107]

In 1810, the African Turf Club was revived and renamed the South African Turf Club, with the arrival of Lord Charles Somerset, as Governor of the Cape Colony, in 1814. Somerset (1767-1831) grew up in Badminton, a centre of English equestrian sports, and was a well known figure at Newmarket. He reorganised the Turf Club, protected the race course,[108] wrested control of the Groote Post Experimental Farm, dedicated to agricultural development, and redirected its focus to horse-breeding, in part to foster the export trade to India, and used the racing industry as a means to improve the breed. He used the rationale that racing was simply a means to an end, a strategy towards improving local horses.[109] This made little sense - draught and transport horses required no Thoroughbred pedigree, indeed draught horses would have benefited from warmblood (solid, weighty draught animal) infusions instead of Thoroughbred - and was likely a mere validation for his personal interest in both racing and breeding Thoroughbreds. As contended by an acerbic observer in 1820:

> [a] distinguished person at the Cape is thought to have conferred a singular benefit on the Colony by bringing over several thoroughbred English horses, and disseminating a taste for racing. But it is difficult to imagine how the interests of an infant Colony can be advanced by the introduction of an animal perfectly useless for any purpose of trade or [agriculture], or by rendering fashionable an expensive and ruinous amusement.[110]

Horses were, however, more than just signifiers of elite status, they could contribute towards *creating* elite status. There was money to be made as a Thoroughbred breeder, which involved an elite group of male breeders from both the Dutch- and English-speaking sectors, including the Governor.[111] Indeed, Somerset relinquished his position after a scandal, in which it appeared as though he had corruptly made a grant of land in consideration of 10,000 *rixdollars*, under the guise of horse trading, and although exonerated, resigned in 1827.[112]

Thoroughbred breeding remained entrenched, and expanded with the expanding racing industry, and the fashion grew during the early and middle nineteenth century for horses with a 'bloodhead', 'light neck' and a 'pedigree'.[113] See, for example, the figure here of Somerset on his horse with a docked tail (in accordance with metropolitan fashion), curved 'Arab' neck, and small 'blood' head considered desirable in an English Thoroughbred.[114] Thus English Thoroughbreds bred for racing and the utilitarian work or farm horse diverged from 1797 onwards, although there remained several points of contact.

Figure 8.2 A View of Lord Charles Somerset (Africana Museum 66/1909).

Conclusion – Closing the Stable Door

Horses played a significant role in social and political processes in the early settlement of the Cape. History was made with horse power and equally the horses were shaped by human history, incorporating the environmental and shifting anthropogenic needs into their genetic makeup. After their hard-fought introduction, resisted by the metropole, horses were first used as draught animals to effect changes in the new environment. Between 1652 and 1662, the utility of the ox was eclipsed in the Cape, particularly in settler perceptions, by the value of the horse, which was utilised first as symbol of power and then as draught animal.

They were also utilised by the VOC authorities as a signifier of difference, a marker of social status. Horses were used to emphasise the difference between native and settler, in order to facilitate the psycho-social subduing of the indigenous population. The role of the horse was predicated on *power*, in both symbolic and material manifestations, and hence became a victim of attack. Horses remained vulnerable in their new environment, threatened by disease and environmental dangers. Unlike in the American settler communities, the horse was not a symbol of wildness but rather of tameness, of civilization and white settlement. There was no basis to re-enact the subduing of the wilderness and the transformation from 'nature' to 'culture' that anthropologists and historians have uncovered, for example, in the western display of rodeo.

The rising number of horses advanced economic and territorial-political ambitions, through military means (on *kommando*) and through the socio-economic control of the horse trade and racing industry. From 1797, two distinct horse cultures emerged – one embraced the British-led racing industry, the other a more utilitarian use of horses. There followed a conflict of horse cultures, between those who followed metropolitan fashions and those adopting 'indigenous' settler modes. Moreover, the divergence led to a morphological difference between the race horses, which were of English Thoroughbred type, and the utilitarian horses, which came to be considered a definite 'breed', known as the Cape Horse. Though the latter were initially not accorded special status, they were later invested with the pride of settler society. Horses, like other livestock breeds in Africa and Southeast Asia have developed in response to a wide range of climates, environmental stresses, and, perhaps particularly, anthropogenic demands.

The 'Ox That Deceives': The Meanings of the 'Basotho Pony' in Southern Africa

Sandra Swart

The men of the tiny mountain kingdom of Lesotho define themselves as 'horsemen'.[1] Indeed, horses are fundamental to mountain life. Horses are used as primary transport over much of the rugged topography of Lesotho, and in some parts used as draught animals, mainly for ploughing the fields. The BaSotho horseman in his traditional blanket is the image which symbolises life in the highlands. Lesotho is small (with under two million people), landlocked, completely surrounded by South Africa, with an history entwined with that of the larger country, and also that of Britain under whose control it was until 1966. Surviving largely on subsistence agriculture, livestock, and remittances from miners employed in South Africa, Lesotho needs tourism as a vital income-generator. The literature on this eco- and cultural-tourism is replete with the romance of the ponies, used by tourists for trekking. The refrains 'the BaSotho are a nation of horsemen', they are 'natural horsemen' and their ponies are 'integral to the landscape' are entrenched – apparent in the popular imagination, repeated in advertisements and travel literature, and even represented on their banknotes. In 1995, when Nelson Mandela paid a state visit to King Moshoeshoe II, he received the honour of the highest order and accepted also another gift from the king, a dappled grey stallion.[2] This gift was given as an icon of the BaSotho people, symbolising their culture and historical identity. Aside from its still primary role in rural transport, particularly in the highlands, the horse has seeped into cultural references, as in the idioms: *Ho ya perere* ('eat the forbidden [horse]', to break a taboo) and *Kepèrè entso* ('he is a black horse', he is strange, maybe dangerous). Horses appear in some myths and have worked themselves into the cosmological weave.[3] They permeate material culture and ritual, as evidenced in, for example, the *Lechoba* – bracelets of horse hair in *mohobelo* men's dance or the popular horse races. The total Basotho Pony population in Lesotho is estimated to be 98,000 to 112,000 animals.

This chapter considers not only the material history of the horse in this small part of what was once the British Empire, but also the social meanings that have come to be encoded in the pony and equally how they play themselves out in terms of 'race' politics. To understand its current significance, the development of the Basotho pony must be traced, exploring its anthropogenic/natural selection as well as colonial scientific breed improvement. Secondly, we must explore the social impact of the animal that is still used today as a symbol of (male) national identity.[4]

People described as 'Sotho' have lived in Southern Africa since the tenth century, but by the sixteenth century, groups had arrived in the region known now as Lesotho, intermingling with the Khoisan (Bushman) people, and establishing small chiefdoms. Widespread trade links were forged between the groups, as well as with external communities.[5] By the early

Map 9.1 Southern Africa. Source: Rev. E. Casalis, *The Basutos; or Twenty-Three Years in South Africa* (London: James Nisbet, 1861).

nineteenth century, there was encroachment from white traders and Boer pioneers, and communities within the territory coming to be known as 'Basutoland' recognised both the external threat and the internal pressure on the environment. Simultaneously, the consolidation and expansion of the neighbouring Zulu state precipitated a violent chain-reaction throughout the sub-region. Those loosely organised southern Sotho groups scattered by Zulu raids were consolidated and amalgamated by King Moshoeshoe, who established a centralised stronghold, and initiated embryonic nation-building.[6]

In 1830 another military force entered the arena occupied by the Sotho: an amorphous ill-defined group, variously identified as the Hottentots, Griquas, 'Bastards', Korannas – with an estimated two thousand men in the field, living largely on plunder, particularly stock theft. They were mounted on 'good horses', had acquired their largely 'oriental', or 'southeast Asian', stock from neighbouring white Cape farmers and owned guns.[7] Some of the first victories over these intruders by the people coming to be called the BaSotho were said to have occurred while a roving band was drunk. The riders were killed and the BaSotho were said to have observed the horses apprehensively until certain they were not the real source from which emanated the explosions of cordite and pain.[8] Even if this story is untrue, it is certain that the BaSotho acquired their first horses by raiding the Kora and Griqua from the early 1830s, and their king systematically acquired them thereafter.[9]

The original SeSotho for horse is *khomo-ea-haka*, 'cattle called Haka' (*Hacqua* being the Khoisan name for a horse).[10] There is, of course, no 'original' word in SeSotho for 'horse', and the word in widespread use by 1950 and in official usage today is *pere*, a close derivative of the Dutch/Afrikaans word '*perd*'.

As noted, Moshoeshoe steadily procured both firearms and horses.[11] Horses remained a precious and vulnerable commodity – there is evidence to suggest that Moshoeshoe tried to suppress stock theft from 1834.[12] Significant evidence of their social value lies in the fact that horses were not private property, their resale was controlled by chiefs and supreme ownership vested in the community. Horses and guns went hand-in-hand. There was rapid widespread use among the BaSotho of the musket, dismounted action with it being supplemented with mounted strikes using axe, *knobkierie* and assegai.[13] Horses and guns became part of the ritual of greeting: 'When within a moderate distance of the camp, the royal attendants alighted from their steeds, fired a salute, then remounted and approached in the rear of their king.'[14] Within a few years there was a mounted force, then a 'nation' of mounted men. Between 1833 and 1838, Moshoeshoe imported 200 horses, and by 1842 was able to tour with a cavalry of 200 mounted men. Horses proved indispensable in repelling mounted raiders and there was a rapid acquisition of stock, soon supplementing acquisition with breeding.[15] By 1861 there was a significant increase; by 1875, the Census states the BaSotho had 35,357 horses. By 1891, they had 81,194. By 1903 every male adult was mounted, although women did not adopt the custom.[16] From 1830 to 1850 imported stock was largely small 'Cape Horses'. In the second wave of acquisitions, from 1835 to 1840, there was predominantly similar stock (with little Thoroughbred blood), while from 1840 to 1870 there was a greater amount of Thoroughbred blood. Horses were not limited to military use – by the latter half of the 1880s had ponies spread to the cultural domain, with pony racing becoming a popular pastime.[17]

As the SeSotho proverb goes, 'All countries are frontiers': the border was porous and the stock acquired was highly various. Imports stopped and exportation began, so that by 1901 the export of horses accounted for 73 per cent of export market value.[18] During the South Africa War (1899-1902) between Britain and the Dutch/Boer Republics, the BaSotho sold their ponies to British remount officers and to the Boer forces.[19] These hardy ponies proved useful: they were of small body size (14-14,2 hands) and able to forage for themselves. This was due to anthropogenic and natural selection: Basutoland (Lesotho) was a harsh environment.[20] It would appear that the BaSotho were good horsemen, but bad horse owners: little shelter, food or veterinary treatment were provided and, added to that, the cold, little forage and precipitous environment led to the development of these small, resilient horses with their famously high endurance.

Significantly, by the early years of the twentieth century there appears to have been a widespread consensus that the Basotho pony had enough individuality to be classified as a 'distinct type'.[21] There were also repeated colonial attempts to improve the breed, through government studs. There were problems after the South African War in encouraging local use of the government stud. It is not an isolated or unprecedented problem. There is a tantalising whisper of early nationalist and identity politics played through the horse as symbol, or indeed as metonym for 'tradition'. There were reports of 'anti-progressive' BaSotho raising an 'outcry' against the efforts of the Government to improve the Basotho pony. They maintained that their ponies were more suitable, more surefooted and enduring, and refused to be part of the breeding programme, contending that they did not need European imports – that the 'indigenous' was better. An official observed the political agenda behind such agricultural programmes as inducement for the BaSotho to see the sense of adopting modern methods; their health and their horseflesh – the latter, after their firearms, their most precious possessions – providing most suitable targets. People who had benefited by the white man's medicine, had seen colts from the Government stallions ... were, at any rate, not indisposed to listen to what white experts in other lines had to tell them.[22]

Between 1903 and 1905, the Resident Commissioner imported seven Arab stallions and offered free stud services. Between 1907 and 1932 mainly Thoroughbred blood was introduced.[23] A mixture of Thoroughbred and Arab studs were widely used between 1900 and 1930. By 1936, R. W. Thornton contended that the once hardy type was utterly debased and required saving – a process that continued sporadically under the Equine Breed Improvement Scheme.[24] By 1966 there were an estimated 90,000 horses in Basutoland, only 10 per cent being of the so-called 'true Basotho type'.

In 1980 the Lesotho government set up the Basotho pony centre, establishing government-run, NGO-funded stud farms.[25] Selection criteria included 'pedigree', 'uniformity of type', together with materially beneficial things like stamina, tripling ability (an ambling lope said to be inherent to 'southeast Asian' stock, discussed in the previous chapter), conformation and fertility.[26] The state established a national stud farm, containing a nucleus herd allocated to a Livestock Improvement Centre, where farmers could bring their mares to be serviced by a 'registered improved stallion'. Secondly, mare camps were set up to control their breeding and prevent unwanted servicing by feral stallions. An equine extension service was offered, focusing on aspects of equine husbandry and nutrition. Endurance and triple racing was promoted as a sport in lieu of simply racing. In 1984, a marketing and trekking centre was also established to use and 'market the Basotho Pony'.

Inventing the Basotho Pony

Yet, the Basotho pony has no written stud-books and no written pedigrees. Although, internationally, it is much better known than any other Southern African 'breed', arguably because of its involvement in the remount trade, there is no historical record of its breeding. This raises questions of autochthony and the creation of historical myth or the 'invention of tradition'. Does the Basotho pony exist? What does it *mean* to say it existed? If between 1870–1890 (its numerical zenith and the high point of its popularity) and 1902 (when the authorities discovered the need to save it), a 'breed' existed (or at least, a particular morphological type), why would one wish to preserve it like a fossil in the amber of that historical moment? Moreover, it must be further asked which admixtures are acceptable: what is the difference between breed *improvement* and breed *submerging*? Equally, there is an internal contradiction. In essence the historical success and survival of the pony type which it is desirable to 'save' is the one that developed over time to be best suited to its environment. However, now there is an insistence on changing previous breeding regimes and a requirement of centralised anthropogenic intervention in order to maintain artificially – indeed, 'bring back' and then 'preserve' – this breed. Why did the natural environment cease producing the kind of stock required?

As discussed in the introduction to this book, the term 'breed' is hard to define. A 'breed' may be understood as a group of animals that, through selection, have come to resemble one another and pass their qualities to their offspring. All this means is that a breed is a population that complies to ancestry. So a 'purebred' animal belongs to an identifiable breed complying with prescribed traits – origin, appearance, and minimum breed standards. As Lush has contended in *The Genetics of Populations* the term is both elusive and subjective:

> [a] breed is a group of domestic animals, termed such by common consent of the breeders ... a term which arose among breeders of livestock, created one might say, for their own use, and no one is warranted in

Figure 9.1 Photograph from the Remount Commission, Basotho Pony, c. 1902.[27]

assigning to this word a scientific definition and in calling the breeders wrong when they deviate from the formulate definition.[28]

So the point at which a collection of animals becomes a 'breed' is a human decision – not a genetic event.[29] This has obvious parallels with human races and their classifications; biological variability exists but it does not conform to racial categories developed by society, largely from the eighteenth century and linked to imperial notions of rule. The reliance on race as a biological concept serves a powerful politico-social purpose by creating an immutable difference between peoples. It is widely accepted that human 'races' are far from natural and are in fact socially produced and shifting, but animal breeds are often a safe realm for those narratives on to which conceptions of human difference such as hierarchy, gender, class, and national character are mapped. Humans cannot be classified accurately into clearly delineated and biologically distinct groups. Particular phenotypic traits are emphasised in typecasting individuals along public views of 'race', a practice which provides little practical information. This is not to state, of course, that there is no biological variation between different populations of people – or indeed horses – just that the biological distinctions are much more opaque than contained in common ideas of 'race'.[30] In the horse world it is simple to form a 'breed': all one needs is paperwork. One registers the 'breed' with the appropriate authorities and promulgates a breed standard. One does not need a lawyer, a threshold number, genetic material or even a horse.

There is also another complication to the colonial history of this idea of 'breed' which opens a further dimension of the politics and paradoxes of animal breeding. In 1950, a committee was set up by the South African Secretary for Agriculture, Dr C. H. Neveling, to investigate indigenous breeds of sheep, cattle, goats and so on that were of interest to Afrikaner culture brokers. This was catalysed by a 1949 letter from a far-flung extension officer, A.W. Lategan, on a distant isolated station in the Northern Cape, who pointed out that, while there was much effort towards soil conservation, indigenous animals well adapted to local conditions were disappearing.[31] The Department of Agriculture subsequently bought twelve representative horses, interestingly not from Lesotho, but from the Orange Free State, and established a breeding station at Nooitgedacht in Ermelo (Mpumalanga) to 'preserve the Basotho pony' and to develop a utility riding horse.[32] It was called the 'Nooitgedacht' pony ostensibly because of the breeders' admixture of Cape Horse/Boerperd and Arab blood to the Basotho base. One may question this rationale, as the Basotho pony was itself developed from Arab and Cape Horse/Boerperd amalgamation. Perhaps a more likely reason was that the South African horse market of the time was reluctant to acquire 'native' stock. A studbook was kept open for females who passed a phenotypic inspection (small Boerperd or Arab). Interestingly, the breeders permitted admixture from Afrikaner breeds, but specifically no British pony breeds were allowed. In 1969 the Breed Society was founded (and changed the name from 'pony' to 'horse' for commercial reasons) with a preference for greys (a trait not linked to Basotho ponies). In 1976, the South African Studbook recognised the Nooitgedacht (invented a mere 24 years before) as its first indigenous breed and sold it at public auctions and gave stock to preserve to the Universities of Pretoria and Stellenbosch. The horses are celebrated as indigenous and autochthonous: 'Thanks to the continued dedication of the breeders, the future of the Nooitgedachter breed is secure. South Africa can justly be proud of this noble and loyal horse',[33] and is part of a 'proud South African heritage'.[34] The Nooitgedacht Horse Breeders' Society states:

> After more than forty years of scientific breeding, dedication and preser-
> vation, the Nooitgedachter is one of South Africa's truly indigenous
> horse breeds, descended from the original Cape horse. Probably one of
> the rarest horse breeds in the world, it is the worthy descendant of the
> Basotho pony, a symbol of tradition and a proud heritage for future
> generations.[35]

This may be analysed in terms of 'hybridity', a term linked with both creol-isation and liminality, with the crossovers of identities generated by colonial-ism.[36] Young reminds us that the hybrid is technically a cross between two different species and that therefore the term hybridisation evokes both the notion of inter-species grafting and the 'vocabulary of the Victorian extreme right', which regarded different races as different species.[37] However, in post-colonial theory, hybridity is intended to suggest all those ways in which this vocabulary was challenged. Even as imperial and racist ideologies insist on racial difference, they catalyse crossovers, partly because not all that happens in the contact zones can be controlled, but occasionally also as a result of deliberate colonial policy. Examples of calculated (colonial) policy used to manipulate cultural outcomes by the means of hybridisation include the 'colonial education policies which aimed to create Europeanised natives, or to use Macaulay's words, a class of persons, Indian in blood and colour, but English in taste, in opinion, in morals and intellect'.[38] Similarly, it may be argued that the Nooitgedacht was *Basotho* in body, *Afrikaans* in identity.

The breed debate reflects the politics of autochthony and the invention of what is 'traditional'. At the turn of the twentieth century, the debate pivoted around the need to import horses to improve the debased, small, degenerate colonial stock. By the mid century mark, and escalating to a peak in 1980, it revolved around the need to preserve or even revive the proud, autochthonous colonial horse for reasons bound up in sentiment, history, nostalgia, heritage, and commerce. This perceived need was so great that NGOs and government funding was directed into programmes in Lesotho, while at the same time an entirely new 'breed' was invented in South Africa in order to preserve the 'heritage' of the Basotho pony. The body of the horse acted as a site of struggle in the protection of national identity and the production of cultural difference.

There is a massive capital invested in the animal breeding industry. Both breeders and the resultant service industries benefit from the public's enthusiasm for 'purebred' animals, preferably registered. There are several parallels to the southern African context in other post-colonial situations, where animals are used as socio-cultural vehicles to promote a sense of self-respect, or – inversely – current cultural ideology is used to market formerly low-priced livestock.[39]

Conclusion

Like the colonisers themselves, the equine 'invaders' were the agents of substantial and enduring changes, in both societal and environmental dimensions. The imperial exchange meant a two-way transformation: in modifying the landscapes, the horse itself underwent a morphological transformation and its function within human society also altered significantly. The horses of empire are bound, connected by blood, history and styles of horsemanship. Southeast Asia and southern Africa are linked together by a common equine heritage: the horse of the Dutch empire in southeast Asia – the Sumbawan, 'Javanese' or 'oriental' stock were transported to the Cape in southern Africa; the stock spread into Basutoland, and horses from both Basutoland and the Cape were relocated to Australia and elsewhere within the Imperial network. War and trade are great diffusers of genotypes. Imperial and later colonial concerns also had unintended effects on trends in demography and disease – both animal and human – and they influenced the way people conceived of elements of nature, such as horses. Equids were pivotal in the economies and societies of the nineteenth century Indian Ocean and South China Sea, crucial in war, trade, transport and leisure. Looking to imperial history and its ecological disruptions, and focusing explicitly on the material contexts of the violence (both physical and figurative) that is connected to the extraction of resources and the alteration of the natural environment, helps explain a process of the establishment of European hegemony in the modern world.[40] What is important for historians to emphasise is the heterogeneity of response: how different native societies within the imperial or colonial network adopted – and adapted to – the horse in very different ways. For example, from the time the Sotho overcame socio-political and economic hurdles and acquired the first horse, the 'ox that deceives', in 1825, they re-invented themselves as an equine society. Horses were not the private property of individuals but rather national assets, with ultimate possession vested in the community. Within a mere two decades, they had transformed Sotho material culture, defence capability, gender relations and social structure. The very sense of what it meant to be an adult Basotho male was transformed: they were *horsemen*. Being deeply embedded in identity politics, equine issues prompted both pride and anxiety.

Declension, or the notion that circumstances are in constant decay, is a prevailing trope in environmental history. Does the history of the horse – even in this circumscribed telling – perhaps offer a new, less declensionist way of thinking about 'invasive' species? There were efforts to contain the horse within the metropolitan grasp – but the horse could not be kept in check. Moreover, although there was a great deal of effort thrown into scientific colonial breed improvement, and ongoing concern over 'deterioration' of type, there was also in several cases the rise of 'indigenous' varieties, invested with the pride of settler or native society. The Basotho pony has become a corporeal icon of nationhood, and, as such, has become embedded in the politics of identity politics and invented traditions. Horses even served to represent an early attempt at liberation from the colonial metropole in the resistance to foreign stallions being used as studs. Perhaps then the horse is not merely an instrument of ecological imperialism, as Crosby suggests, but also of ecological resistance?

EPILOGUE

10 'Together yet Apart': Towards a Horse-story

Greg Bankoff and Sandra Swart

The horse may seem the least likely of all animals to be the subject of historical 'othering' given its close and enduring relationship to humanity. While its central role has received some attention in European history writing to the extent that the development of feudalism has even been traced to the invention of the horse-stirrup, 'equine orientalism' persists in the non-western-world, familiarity breeding if not contempt then at least neglect.[1] Non-western societies may have received increasing attention in recent decades, reflecting not only a greater awareness of the importance in understanding the wider historical experience of the majority of the world's population but also a shift in the locus of gravity away from the North to the East and South, but still very little scholarship aims at exploring inter-regional perspectives outside of the European and North American context. The norm is still very much a focus on intra-metropole or metropole-colonial relationships, ensuring that modern historiography 'continues to run along channels excavated by colonial discourse'.[2]

These historiographical channels perpetuate the orientalisation of animals in disallowing non-human creatures their own history. Horses appear as objects to be fought over by people, as instruments to help conquer people and lands, essentially as backdrops or props in the unfolding human drama. This trend has a parallel in the historiography of human social relations. Historians of the world in 1400, on the brink of European expansion, have now critiqued the way the metropolitan-peripheral dichotomy has been set up by previous scholars. Recent studies have sought to show instead that the expansion of European societies not only affected the colonised, but also dramatically altered those European societies themselves. For example, Eric Wolf's *The People Without History* adopted the Marx and Engels phrase to convey a deep irony. The phrase refers to those peoples who lack a formally written articulation of their histories, and caustically suggests that quite the reverse was true: that the peoples encountered by the European metropolitans' expansionist projects were actually characterised by their own long and complex histories, and, moreover, were generally not isolated but interconnected. A similar irony might pervade future writings on horses and humans. Equally, by emphasising a common past, less anthropocentric studies could escape the polarities of human and horse, just as Wolf has urged for the false binary of active 'white' centre and passive 'non-white' periphery. The horse, of course, has its own history quite apart from that of humanity though the two have run along parallel and converging lines since its domestication some five thousand years ago.[3] There is no denying the increasing centrality of this relationship but neither should it obscure the separateness of equine history: it is a reciprocal partnership though one where the terms of exchange favour human over equid much as they do in any patron–client relationship. The difficulty is just as much how to express this 'together yet apart' as it is to counter Erica Fudge's 'present yet absent' in history.

By focusing on the horse in the Indonesian archipelago and Thailand, however, and its expansion into the Philippines and southern Africa, another trajectory starts to suggest itself. This scrutiny circumvents the distorting mirror of imperial historiography, and draws the eye instead to South–South comparisons, while seeing the role of the metropole more realistically. These regions may have been brought into greater economic, cultural and biological proximity quite inadvertently in the early modern period as a result of the colonial venture but the genetic transfer of equine genes and phenotypes binds their subsequent histories together much as in studies of the Irish or Jewish diasporas. A number of intriguing paths of study present themselves: What different effects did the introduction of the horse have on existing ecosystems? How did the horse adapt relative to its new environment? In what ways did the horse interact with indigenous human cultures? Were these distinct from the way 'colonial' societies evolved? The present study begins to address only some of these questions. It hints at the outline of a more *species-centric* history to counterpoise against that of the prevalent anthropocentric one: a *horse*-story to lie alongside that of a *her*-story or a *his*-story. It also contributes to a wider sense of colonial narrative, a more inclusive form of historiography that explains many historical events in terms of the 'biotic interactions' between people, animals and land.[4]

Notes

Chapter 1. Breeds of Empire and the 'Invention' of the Horse
Greg Bankoff and Sandra Swart

1   Nicholas Russell, *Like Engend'ring Like: Heredity and Animal Breeding in Early Modern England*, London: Cambridge University Press, 1986, p. 23.

2   M. S. Drewer, 'The Domestication of the Horse'. In Peter Ucko and George Dimbleby (eds), *The Domestication and Exploitation of Plants and Animals*, London: Duckworth, 1969, pp. 471–478; and Robin Law, *The Horse in West African History: The Role of the Horse in the Societies of Pre-colonial West Africa*, Oxford: Oxford University Press, 1980, pp. 1–3.

3   Despite the animal's initial evolution in the Americas, the horse had become extinct there some 13,000 years ago and was not reintroduced until the arrival of the Spaniards in the late fifteen and early sixteenth centuries. Tim Flannery, *The Eternal Frontier: An Ecological History of North America and Its Peoples*, London: William Heinemann, 2001, pp. 111–112, 293–296. The first horse to set foot in the Americas was carried on Christopher Columbus's second voyage and arrived on the Caribbean island of Hispañola in 1493. Hernan Cortés subsequently took 15 with him on his conquest of Mexico in 1519. On horses in America, see Virginia DeJohn Anderson, *Creatures of Empire: How Domestic Animals Transformed Early America*, New York: Oxford University Press, 2004; and in Australia, see Alexander T. Yarwood, *Walers: Australian Horses Abroad*, Melbourne: Melbourne University Press, 1989.

4   On historical methods of horse breeding, see Russell, *Like Engend'ring Like*, 1986, pp. 58–121; and on modern breeding efforts, see Margaret Derry, *Bred for Perfection: Shorthorn Cattle, Collies, and Arabian Horses since 1800*, Baltimore: Johns Hopkins University Press, 2003; and Tim Birkhead, *A Brand-New Bird: How Two Amateur Scientists Created the First Genetically Engineered Animal*, New York: Basic Books, 2003.

5   Alfred Crosby, *The Columbian Exchange: Biological Consequences of 1492*, Westport: Greenwoods, 1972.

6   Claude Lévi-Strauss, *Totemism*, Boston: Beacon Press, 1963, p. 89.

7   Alfred Crosby, *Ecological Imperialism: The Biological Expansion of Europe, 900–1900*, New York: Cambridge University Press, 1986, p. 194.

8   For an ovine comparison, or the 'ungulate irruption', see Elinor Melville, *A Plague of Sheep: Environmental Consequences of the Conquest of Mexico*, Cambridge: Cambridge University Press, 1994.

9   It has also been referred to as the 'animal turn' or 'animal studies'. There are currently at least four major journals that publish in the field of animal-related topics: *Anthrozöos, Animal Welfare, Journal of Animal Ecology* and *Society and Animals*. There is also a growing literature on animal histories like Harriet Ritvo, *The Animal Estate: The English and Other Creatures in the Victorian Age*, Cambridge: Harvard University Press, 1987, as well as such 'standards' like Keith Thomas, *Man and the Natural World*, New York: Pantheon Books, 1983 and James Serpell, *In the Company of Animals*: A study of Human–Animal Relationships, Oxford: Blackwell, 1986.

10 See, for example, conferences like 'Animals in History and Culture' at Bath Spa University College, 2000; 'Representing Animals' at the University of Wisconsin–Milwaukee, 2000; 'Thresholds of Identity in Human–Animal Relationships' at the University of California, 2000; and 'Millennial Animals: Theorising and Understanding the Importance of Animals' at the University of Sheffield, 2000.

11 See, for example, Jonathan Burt, *Animals in Film*, London: Reaktion Books, 2002; Eileen Crist, *Images of Animals: Anthropomorphism and Animal Mind*, Philadelphia: Temple University Press, 1999; Ritvo, The *Animal Estate*, 1987; Nigel Rothfels (ed.), *Representing Animals*, Bloomington: University of Indiana Press, 2003; Chris Philo and Chris Wilbert (eds), *Animal Spaces, Beastly Places: New Geographies of Human–Animal Relations*, London: Routledge, 2000.

12 Chris Philo, 'Through the Geographical Looking Glass: Space, Place, and Society-Animal Relations', *Society and Animals*, vol. 6, no. 2, 1998. See also see Harriet Ritvo, 'History and Animal Studies', *Society and Animals*, vol. 10, 2002, pp. 403–406; Steve Baker, *Picturing the Beast: Animals, Identity and Representation*, Manchester: Manchester University Press, 1993; Adrian Franklin, *Animals and Modern Cultures: A Sociology of Human–Animal Relations in Modernity*, London: Sage, 1999; Mary Henninger-Voss, *Animals in Human Histories: The Mirror of Nature and Culture*, Rochester: University of Rochester Press, 2002; Kathleen Kete, *The Beast in the Boudoir: Pet-keeping in Nineteenth Century Paris*, Berkeley: University of California Press, 1994; Aubrey Manning and James Serpell, *Animals and Human Society, Changing Perspectives*, London: Routledge, 1994; Robert Mitchell, Nicholas Thompson and H. Lyn Miles (eds) *Anthropomorphism, Anecdotes and Animals*, Albany: State University of New York Press, 1997; Ritvo, *The Animal Estate*, 1987; Joyce Salisbury, *The Beast Within: Animals in the Middle Ages*, New York:

Routledge, 1986; Serpell, *In the Company of Animals*, 1986; Joanna Swabe, *Animals, Disease and Human Society: Human-Animal Relations and the Rise of Veterinary Medicine*, London: Routledge, 1999; Thomas, *Man and the Natural World*, 1983; Yi-Fu Tuan, *Dominance and Affection: The Making of Pets*, New Haven: Yale University Press, 1984; and Roy Willis (ed.), *Signifying Animals: Human Meaning in the Natural World*, London: Routledge, 1990.

13 Anderson, 'A Walk on the Wild Side:' 1997, pp. 463–485; Charles Bennett, 'Cultural Animal Geography: An Inviting Field of Research', *Professional Geographer*, vol. 12, no. 5, 1960, pp. 12–14; and Jennifer Wolch, and Jody Emel (eds), *Animal Geographies: Place, Politics and Identity in the Nature–Culture Borderlands*, London: Verso, 1998.

14 See the discussion by R. Yarwood, N. Evans, and J. Higginbottom, 'The Contemporary Geography of Indigenous Irish Livestock', *Irish Geography*, vol. 30, 1997, pp. 17–30, especially p. 17.

15 Wolch and Emel, *Animal Geographies*, 1998, p. xi.

16 Historical works such as Juliet Clutton-Brock, *A Natural History of Domesticated Mammals*, Cambridge: Cambridge University Press, 1989 and Russell, *Like Engend'ring Like*, 1986 offer clear, scientifically informed accounts of complex processes and relationships.

17 Some historians have used animal artefacts to reconstruct pet-keeping relations. See: Nancy Carlile, 'The Chewed Chair Leg and the Empty Collar: Mementos of Pet Ownership in New England', *Dublin Seminar for New England Folklife Annual Proceedings*, vol. 18, 1993, pp. 130–146; and Neil Dana Glukin 'Pet Birds and Cages of the 18th Century', *Early American Life*, vol. 8, 1977, pp. 38–41, 59.

18 See, for example, Nicolas Rupke (ed.), *Vivisection in Historical Perspective*, London: Routledge, 1987; and Richard French, *Antivivisection and Medical Science in Victorian Society*, Princeton: Princeton University Press, 1975.

19 Susan Jones, *Valuing Animals: Veterinarians and Their Patients in Modern America*, Baltimore: Johns Hopkins University Press, 2003. For further reflections on this topic, see Ritvo, 'History and Animal Studies', 2002, pp. 403–406. This issue of *Society and Animals* also includes essays on the relation of animal studies to other disciplines in the humanities and social sciences. See also the conference on 'The Chicken: Its Biological, Social, Cultural, and Industrial History, from the Neolithic Middens to McNuggets' sponsored by the Yale Program in Agrarian Studies in 2002.

20 Wolch and Emel, *Animal Geographies*, 1998, p. xi.

21 This position is not without critique. Many critics dismiss animal activism as a further commodification of animals rather than a break from it. See Neil Smith, 'The Production of Nature'. In George Robertson, Melinda Mash, Lisa Tickner, Jon Bird, Barry Curtis, and Tim Putnam (eds), *Future Natural*

Nature/Science/Culture, London and New York: Routledge, 1996, pp. 35–54; and Wolch and Emel, *Animal Geographies*, 1998, p. xii. Equally, many of those on the left fear that the concern may in fact be premised on a return to pre-modern 'animistic' beliefs.

22 Baker, *Picturing the Beast*, 2001, p. xvi. Although, he adds, animals themselves cannot be discussed, only their representations.

23 Molly Mullin, 'Mirrors and Windows: Sociocultural Studies of Human–Animal Relationships', *Annual Review of Anthropology*, vol. 28, 1999, pp. 201–224.

24 Barbra Noske, Humans and Other Animals: Beyond the Boundaries of Anthropology, Pluto Press, 1989; A. N. Rowan, 'The Human–Animal Interface: Chasm or Continuum?' In Michael Robinson and Lionel Tiger (eds), *Man and Beast Revisited*, Washington: Smithsonian Institution Press, 1991.

25 Peter Boomgaard, *Frontiers of Fear: Tigers and People in the Malay World 1600–1950*, New Haven: Yale University Press, 2001.

26 For examples of integrated texts on a macro-level, see William McNeill and John R. McNeill, *The Human Web: A Bird's-Eye View of World History*, New York: W.W. Norton, 2003; Crosby, *Ecological Imperialism*, 1986; and Jared Diamond, *Guns, Germs, and Steel: The Fates of Human Societies*, New York: W.W. Norton, 1997. For pioneer texts on micro-level, see James Turner, *Reckoning with the Beast: Animals, Pain and Humanity in the Victorian Mind*, Baltimore: John Hopkins University Press, 1980; Thomas, *Man and the Natural World*, 1983; Ritvo *The Animal Estate*, 1987; and Kete, *The Beast in the Boudoir*, 1994.

27 Wolch and Emel, *Animal Geographies*, 1998, p. xvii.

28 Roy Willis, *Man and Beast*, New York: Basic Books, 1974; and Claude Lévi-Strauss, *The Savage Mind*, Chicago: Chicago University Press, 1967.

29 Paul Shepard, *The Others: How Animals Made Us Human*, Washington, D.C.: Island Press, 1995.

30 Chris Philo, 'Animals, Geography, and the City: Notes on Inclusions and Exclusions'. In Wolch and Emel, *Animal Geographies*, 1998, pp. 51–71.

31 Kay Anderson, 'Animals, Science, and Spectacle in the City'. In Wolch, and Emel, *Animal Geographies*, 1998, pp. 27–50.

32 Jody Emel, 'Are You Man Enough, Big and Bad Enough? Wolf Eradication in the US'. In Wolch and Emel, *Animal Geographies*, 1998, pp. 91–116.

33 Raymond Firth, *Elements of Social Organization*, London: Watts, 1951; J. Berger, "Animal World", *New Society*, 1971, p. 1043; Mary Douglas, 'Deciphering A Meal'. In Clifford Geertz (ed.), *Myth, Symbol and Culture*, New York: W. W. Norton, 1971; Clifford Geertz, 'Deep Play Notes on the Balinese Cockfight'. In Clifford Geertz (ed.), *Myth, Symbol and Culture*, 1971, p. 138; Joseph Klaits and Barrie Klaits (eds) *Animals and Man in Historical Perspective*. New York: Harper & Row, 1974.

34 A model example is Marvin Harris's ecological explanation of the sacred cows of the Hindu. His point is that any large human population will always make sure it gets as much energy as possible out of the technological and ecological situation in which it is placed. Religions and ideologies are adaptations to the techno-ecological system's needs. In this manner the practice of the sacred cows of the Hindu is not irrational because the energetic advantage of using the cows as draught animals, using their dung as fuel and fertilisers and consuming the milk, is bigger than butchering them for food. Marvin Harris, 'The Myth of the Sacred Cow', *Man, Culture and Animals*, 1965, p. 225.

35 Claude Lévi-Strauss, *The Savage Mind*, 1966, p. 104. Another influential structuralist or symbolic approach is found in an influential work by Edmund Leach (1964) where he connected the use of animals in insults to dietary and sexual prohibitions. For instance the pig is used frequently in verbal abuse: 'You swine! You filthy pig!' Pigs are used in insults like these, suggests Leach, because there is a certain guilt connected to our use of pigs: we rear them for the sole purpose of eating them (while sheep provide wool, cows milk, etc.). This is a rather shameful thing, a shame that quickly attaches to the pig itself. Edmund Leach, 'Anthropological Aspects of Language: Animal Categories and Verbal Abuse', in E. Lenneberg (ed), *New Directions in the Study of Language*, Boston: MIT Press, 1964.

36 Harriet Ritvo, 'Animal Planet', Environmental History, vol. 9, no. 2, 2004, p. 204.

37 Erica Fudge, *Perceiving Animals: Humans and Beasts in Early Modern English Culture*, Basingstoke: Macmillan Press, 2000, p.3.

38 Edward Said, *Orientalism*, New York: Pantheon Books, 1978.

39 Bernard Williams, 'Prologue: Making Sense of Humanity'. In James Sheehan and Morton Sosna (ed.), *The Boundaries of Humanity: Humans, Animals, Machines*, Berkeley and Los Angeles: University of California Press, 1991, pp. 13–23; and Jack Goody, *The Power of Written Tradition*, Washington: Smithsonian Institute Press, 2000. According to the French philosopher René Descartes, animals are confined by instinctual sensations that are purely bodily or mechanical in nature and are not ones made in consciousness or embodied by the intellect. Mary Midgley, *Animals and Why They Matter*, Athens: Georgia University Press, 1984.

40 Jeffrey Mason and Susan McCarthy, *When Elephants Weep: The Emotional Lives of Animals*, New York: Delta, 1995, pp. 32–35.

41 Post-modernists do not deny that history has narrative: only that the narrative should be singular. On narrative and history, see William Cronon, 'A Place for Stories: Nature, History, and Narrative', *Journal of American History*, vol. 78, no. 2, 1992, pp. 1347-1376; and Hayden White, *The Content*

of the Form: Narrative Discourse and Historical Representation, Baltimore: Johns Hopkins University Press, 1987.

42 Charles Darwin, *The Expression of the Emotions of Man and Animals*, London: Murray, 1872.

43 George Romanes, *Animal Intelligence*, New York: Appleton, 1882; and George Romanes, *Mental Evolution in Animals*, London: Kegan, Paul, Trench, 1883.

44 Edward Thorndike, 'Animal Intelligence: An Experimental Study of the Associative Processes in Animals', *Psychological Monographs*, vol. 2, no. 4 and 8, 1898; and Edward Thorndike, *Animal Intelligence: Experimental Studies*, New York: Macmillan, 1911.

45 J. Stam Henderikus and Tanya Kalmanovitch, 'E. L. Thorndike and the Origins of Animal Psychology; On the Nature of the Animal in Psychology', *American Psychologist*, vol. 53, no. 10, 1998, pp. 1138-1142.

46 Roger Lewin, 'I Buzz Therefore I Think', *New Scientist*, 15 January, 1994, p.3; and Masson and McCarthy, *When Elephants Weep*, 1995.

47 The late John Kennedy, professor of animal behaviour at Imperial College, London, maintained that no matter how adaptive or complex an animal's action may appear, it was still the result of the power of natural selection to optimise behaviour and was not proof of its consciousness. John Kennedy, *The New Anthropomorphism*, Cambridge: Cambridge University Press, 1992.

48 Marian Dawkins, *Through Our Eyes Only? The Search for Animal Consciousness*, Oxford: W. H. Freeman Spektrum, 1993, p. 14. The focus of her work is mainly on chickens.

49 Shelia Rowbotham, *Hidden from History: 300 Years of Women's Oppression and the Fight Against It*, London: Pluto Press, 1973.

50 Such ideological strictures on women were never fully accepted and women have struggled in many western societies 'to overturn their identification with the home as conceived of as a space of privacy, stasis, tradition and connectedness'. Kay Anderson, 'A Walk on the Wild Side', 1997, p. 476.

51 Gisela Blok, 'Challenging Dichotomies: Perspectives on Women's History'. In Karen Offen, Ruth Roach Pierson and Jane Rendall (eds), *Writing Women's History, International Perspectives*, Basingstoke, Hampshire: Macmillan, 1991, p. 2.

52 Joan Kelly, *Women, History and Theory: The Essays of Joan Kelly*, Chicago and London: University of Chicago Press, 1984, p. 1. The essay was first published in *Signs, The Journal of Women in Culture and Society*, vol. 1, no. 4, 1976, pp. 809-823.

53 Donna Haraway, *Primate Visions: Gender, Race, and Nature in the World of Modern Science*, New York and London: Routledge, 1989, pp. 10-13. Haraway's argument is concerned with primates but applies to a lesser extent

to other animals. On the animal–human divide, see Lévi-Strauss, *Totemism*, 1963 on totemism; Rane Willerslev, 'Not Animal, Not Not-Animal: Hunting, Imitation and Empathetic Knowledge among the Siberian Yukaghirs', *Journal of the Royal Anthropological Institute*, vol. 10, 2004, pp. 629–652 on hunting; and Eduardo Viveiros de Castro, *From the Enemy's Point of View: Humanity and Divinity in an Amazonian Society*, Chicago: University of Chicago Press, 1992 on cannibalism.

54 Joan Thirsk, *Horses in Early Modern England For Service, For Pleasure, For Power*, Reading: University of Reading, 1978; Robert Delort, *Les Animaux ont une Histoire*, Paris: Éditions du Seuil, 1984; Thomas, *Man and the Natural World*, 1984; Crosby, *Ecological Imperialism*, 1986; Ritvo, *The Animal Estate*, 1987; and Robert Malcolmson and Stephanos Mastoris, *The English Pig: A History*, London and Rio Grande: Hambledon Press, 1998. The term neo-Europe is used after the manner employed by Alfred Crosby to signify the settler societies of North America, South America and Australasia. The South African case is more ambivalent. On South Africa, see John Richards, *The Unending Frontier: An Environmental History of the Early Modern World*, Berkeley: University of California Press, 2003, pp. 274–306.

55 Briton Busch, *War Against the Seals: A History of the North-American Seal Fishery*, Kingston and Montreal: McGill-Queen's University Press, 1985; Arthur McEvoy, *The Fisherman's Problem: Ecology and the Law in the California Fisheries 1850–1980*, Cambridge: Cambridge University Press, 1986; Ritvo, *The Animal Estate*, 1987, pp. 205–88; Matt Cartmill, *A View to a Death in the Morning: Hunting and Nature through History*, Cambridge: Harvard University Press, 1993; Andrew Isenberg, *The Destruction of the Bison*, Cambridge: Cambridge University Press, 2000; and Nigel Rothfels, *Savages and Beasts: The Birth of the Modern Zoo*, Baltimore and London: Johns Hopkins University Press, 2002.

56 Richard Bulliet, *The Camel and the Wheel*, Cambridge: Harvard University Press, 1975; Law, *The Horse in West African History*, 1980; Boomgaard, *Frontiers of Fear*, 2001; Lance van Sittert and Sandra Swart (eds), 'Canis Familiaris – A Dog History of South Africa', *South African Historical Journal*, vol. 48, 2003, pp. 138–251; and John Knight (ed.), *Wildlife in Asia: Cultural Perspectives*, London and New York: Routledge Curzon, 2004. Tim Ingold's work on reindeer ranching in Lapland, which though geographically part of Europe is atypical of the region, ranges more widely across the Arctic tundra and taiga. Tim Ingold, *Hunters, Pastoralists and Ranchers: Reindeer Economies and Their Transformations*, Cambridge: Cambridge University Press, 1980. See also Olga Crisp's study on the labour demand for horses on an estate in Central Russia in the late nineteenth century (1983). Olga Crisp, 'Horses and Management of a Large Agricultural Estate in Russia at the End of the Nineteenth Century'. In Francis Thompson (ed.), *Horses in European Economic History: A Preliminary*

Canter, Reading: The British Agricultural History Society, 1983, pp. 156-176. These lists, of course, are far from all inclusive.

57 Crosby, *The Columbian Exchange*, 1972.

58 Flannery, *The Eternal Frontier*, 2001.

59 Hans Zinsser, *Rats, Lice, and History: Being a Study in Biography, which, after Twelve Preliminary Chapters Indispensable for the Preparation of the Lay Reader, Deals with the Life History of Typhus Fever*, London: Papermac, 1963 (first published 1935), p. 201.

60 Crosby, *The Columbian Exchange*, 1972; and Crosby, *Ecological Imperialism*, 1986.

61 Melville, *A Plague of Sheep*, 1994.

62 Ibid., pp. 6-7.

63 Ritvo, *The Animal Estate*, 1987, p. 4.

64 John Berger, 'Why look at animals?' In John Berger, *About Looking*, New York: Pantheon Books, 1980, p. 2.

65 See, for example, Steve Baker, *Picturing the Beast*, 2001.

66 For accounts of early breeding, see Russell, *Like Engend'ring Like*, 1986; Harriet Ritvo, 'Possessing Mother Nature: Genetic Capital in 18th-Century Britain'. In Susan Staves and John Brewer (eds), *Early Modern Conceptions of Property*, London: Routledge, 1994, pp. 413-426; and Ritvo, *The Animal Estate*, 1987, Chapter 2. Also, see fn. 4 above.

67 Jay L. Lush, *The Genetics of Populations*, Ames: Iowa Agriculture and Home Economics Experiment Station, College of Agriculture, Iowa State University, 1994.

68 Clutton-Brock, *A Natural History of Domesticated Mammals*, 1989.

69 Jared Diamond, *Guns, Germs and Steel*, 1997, p. 159.

70 For an extended discussion see Harriet Ritvo, *The Platypus and the Mermaid, and Other Figments of the Classifying Imagination*, Cambridge: Harvard University Press, 1997.

71 When horse racing became organised through the newly formed Jockey Club (established 1752), a flood of equine portraits resulted, which increased with the arrival of artists like George Stubbs and Sawrey Gilpin. There was a popular demand by the affluent to be painted with his favorite hunter or race horse.

72 Ritvo, *The Animal Estate*, 1987, pp. 1-3.

73 Donna Landry, 'The Bloody Shouldered Arabian and Early Modern English Culture'. *Criticism*, Winter, 2004.

74 However, 'blood tells' remains a popular idea in the racing world even today, applicable to people as well as horses. See Rebecca Cassidy, *The Sport of*

Kings: Kinship, Class and Thoroughbred Breeding in Newmarket, Cambridge: Cambridge University Press, 2002, pp. 140-160.

75 Joan Thirsk (ed.), *The Agrarian History of England and Wales*, Volume VII, *1640-1750: Agrarian Change*, Cambridge: Cambridge University Press, 1985, p. 578.

76 Peter Edwards, *The Horse Trade of Tudor and Stuart England*, Cambridge: Cambridge University Press, 1988, p. 51.

77 Thirsk, *Horses in Early Modern England*, 1978, p. 578.

78 Swart, Chapter 8, this volume.

79 Works on horses are few. See Bernice de Jong Boers 'Paardenfokkerij op Sumbawa (1500-1930)' [Horse Breeding in Sumbawa], *Spiegel Historiael*, vol. 10/11, no. 32, 1997, pp. 438-443 and more recently the important section on livestock in Peter Boomgaard and David Henley (eds), *Smallholders and Stockbreeders: Histories of Foodcrop and Livestock Farming in Southeast Asia*, Leiden: KITLV Press, 2004. There is also Roderich Ptak's study of the horse trade to China that includes Southeast Asia. Roderich Ptak, 'Pferde auf See: ein vergessener Aspekt des maritimen chinesischen Handels im frühen 15. Jahrhundert'. In Roderich Ptak, *China's Seabourne Trade with South and Southeast Asia (1200-1750)*, Aldershot: Ashgate, 1999, pp. 199-233; and William Gervase Clarence-Smith's fairly recent work on the economic role of Arabs in horse trading. William Gervase Clarence-Smith, 'Horse Trading: The Economic Role of Arabs in the Lesser Sunda Islands, c. 1800 to c. 1940'. In Huub de Jonge and Nico Kaptein (eds), *Transcending Borders: Arabs, Politics, Trade and Islam in Southeast Asia*, Leiden: KITLV Press, 2002, pp. 143-162.

80 On horses in the mainland, see William Gervase Clarence-Smith, 'Horse Breeding in Mainland Southeast Asia'. In Boomgaard and Henley, *Smallholders and Stockbreeders*, 2004, pp. 191-192. On the use of horses in warfare, see Michael Charney, *Southeast Asian Warfare, 1300-1900*, Leiden: Brill, 2004.

81 Jan Wisseman Christie, 'The Agricultural Economies of Early Java and Bali'. In Boomgaard and Henley (eds), *Smallholders and Stockbreeders: Histories of Foodcrop and Livestock Farming in Southeast Asia*, Leiden: KITLV Press , 2004, pp. 47-68; and Bernice de Jong Boers, Chapter 4, this volume.

82 Greg Bankoff, Chapter 6, this volume.

83 Peter Boomgaard, Chapter 3, this volume; and de Jong Boers, Chapter 4, this volume.

84 The definition of a pony is a horse under 14-and-a-half hands or 147 centimetres.

85 William Gervase Clarence-Smith, Chapter 2, this volume; and Dhiravat na Pomberja, Chapter 5, this volume.

86 Bankoff, chapters 6-7, this volume; and Sandra Swart, chapters 8-9, this volume.

87 Sandra Swart, 'Riding High – Horses, Power and Settler Society, c. 1654-1840', Kronos, vol. 29, (Environmental History Special Issue), November 2003; Sandra Swart, '"Race" Horses – A Discussion of Horses and Social Dynamics in Post-Apartheid Southern Africa'. In N. Distiller and M. Steyn (eds), *Under Construction: 'Race' and Identity in South Africa Today*, Sandton: Heinemann, 2004, pp. 13-24; and Sandra Swart, ' "Horses! Give Me More Horses!" – White Settler Society and the Role of Horses in the Making of Early Modern South Africa'. In Karen Raber and Treva Tucker (eds), *The Culture of the Horse: Status, Discipline, and Identity in the Early Modern World*, Basingstoke: Palgrave Macmillan, 2005, pp. 311-328.

88 For a discussion see Thomas Dunlap, *Nature and the English Diaspora – Environment and History in the United States, Canada, Australia and New Zealand*, Cambridge: Cambridge University Press, 1999.

89 One hand measures 4 inches or 10.16 cm.

90 Swart, Chapter 9, this volume.

Chapter 2. Southeast Asia and Southern Africa in the Maritime Horse Trade of the Indian Ocean, c. 1800-1914
William Gervase Clarence-Smith

1 William G. Clarence-Smith, 'Cape to Siberia: The Indian Ocean and China Sea Trade in Equids'. In David Killingray, Margaret Lincoln and Nigel Rigby (eds), *Maritime Empires: British Imperial Maritime Trade in the Nineteenth Century*, Woodbridge: Boydell and Brewer, 2004, pp. 48-67. My thanks are due to the editors for permission to draw heavily on this chapter for the present essay.

2 Alexander T. Yarwood, *Walers: Australian Horses Abroad*, Melbourne: Melbourne University Press, 1989.

3 J. L. Kipling, *Beast and Man in India, a Popular Sketch of Indian Animals in their Relation with People*, London: Macmillan, 1921.

4 B. R. Mitchell, *International Historical Statistics*, London: Macmillan, 1998; *Statesman's Year-Book*, London: Macmillan, various years; *Yearbook of Food and Agricultural Statistics*, Washington: Food and Agriculture Organization, 1947. Some figures have been extrapolated from later data.

5 Roderich Ptak, 'Pferde auf See, Ein Vergessener Aspekt des Maritimen Chinesischen Handels im Frühen 15. Jahrhundert', *Journal of the Economic and Social History of the Orient*, vol. 34, no. 2, 1991, pp. 199-233.

6 H. Epstein, *The Origin of the Domestic Animals of Africa*, New York: Africana, 1971, vol. 2, pp. 474-476; G. Tylden, *Horses and Saddlery*, London: J. A. Allen & Company, 1965, pp. 58-59. See Sandra Swart, Chapter 8, this volume.

7   *Official Year Book of the Union, and of Basutoland, Bechuanaland Protector-ate, and Swaziland,* Government Printer: Pretoria, vol. 5, 1922, p. 495. See Sandra Swart, Chapter 9, this volume.

8   Eric Rosenthal, *Encyclopaedia of Southern Africa,* London: Warne, 1964, pp. 234-235; R. S. Summerhays, *The Observer's Book of Horses and Ponies,* London: Frederick Warne & Co., 1954, pp. 63-64.

9   Jane Kidd, *The Horse: The Complete Guide to Horse Breeds and Breeding,* London: Longmeadow Press, 1985, pp. 14, 90, 93, 190; Summerhays, *The Observer's Book,* 1954, pp. 18, 54-55; Yarwood, *Walers,* 1989.

10 Clarence-Smith, 'Cape to Siberia', 2004; and Yarwood, *Walers,* 1989.

11 Jos Gommans, 'The Horse Trade in Eighteenth-century South Asia', *Journal of the Economic and Social History of the Orient,* vol. 37, 1994; Hala M. Fattah, *The Politics of Regional Trade in Iraq, Arabia and the Gulf, 1745-1900,* Albany (New York): State University of New York Press, 1997.

12 C. A. Bayly, *Indian Society and the Making of the British Empire,* Cambridge: Cambridge University Press, 1990, pp. 142-143; Gommans, 'The Horse Trade', 1994, pp. 241-247.

13 G. Tylden, *Horses and Saddlery: An Account of the Animals used by the British and Commonwealth Armies from the Seventeenth Century to the Present Day, with a Description of their Equipment,* London: J. A. Allen, 1980 (reprint of 1965 edn), p. 58.

14 Marcus Arkin, *Storm in a Teacup: The Later Years of John Company at the Cape, 1815-36,* Cape Town: Struik, 1973, pp. 212-214, 228-230.

15 Yarwood, *Walers,* 1989, pp. 2, 30-31; Tylden, *Horses and Saddlery,* 1980, pp. 52-53, 58.

16 George M. Theal, *Records of the Cape Colony,* London: Clowes, 1897-1905, vol. 17, pp. 488-491; vol. 32, p. 478.

17 Edward Balfour, *Cyclopaedia of India and of Eastern and Southern Asia, Commercial, Industrial and Scientific,* Madras, 1871, 2nd edn, vol. 2, p. 618.

18 *Parliamentary Papers,* Trade Reports.

19 British Library, Oriental and India Office Collections, Vanrenen Collection, Adrian Vanrenen 24 August 1868 and 10 February 1870, and Jacob Vanrenen 15 March 1869; Tylden, *Horses and Saddlery,* 1980, pp. 59-60; Yarwood, *Walers,* pp. 27, 52. One hand equals four inches.

20 Yarwood, *Walers,* 1989, p. 79; *Official Year Book,* 1922, p. 495.

21 Robert Wallace, *Farming Industries of Cape Colony,* London: King, 1896, pp. 309-310; Yarwood, *Walers,* 1989, pp. 122-123.

22 Yarwood, *Walers,* 1989, pp. 82-83, 123; C. E. Callwell, *Small Wars: a Tactical Textbook for Imperial Soldiers,* Novato: Presidio Press, 1990 (reprint of 1906 edn), pp. 402-409.

23 *Official Year Book*, 1922, p. 497.

24 Yarwood, *Walers*, 1989, p. 102.

25 Wallace, *Farming Industries*, 1896, pp. 308, 316-319, 322; British South Africa Company, *Reports on the Administration of Rhodesia, 1898-1900*, London, 1900, pp. 202-205.

26 Mitchell, *International Historical Statistics*, 1998, Table C11.

27 Epstein, *The Origin*, vol. 2, 1971, p. 477; Yarwood, *Walers*, 1989, p. 123.

28 Yarwood, *Walers*, 1989, p. 102; Tylden, *Horses and Saddlery*, 1980, p. 50.

29 Somerset Playne, *Cape Colony (Cape Province): Its History, Commerce, Industries and Resources*, London: Foreign and Colonial Compiling and Publishing Co., 1910-1911, p. 619; Tylden, *Horses and Saddlery*, 1980, p. 60.

30 Wallace, *Farming Industries*, 1896, pp. 314, 319.

31 George Blake, *B. I. Centenary, 1856-1956*, London: Collins, 1956, pp. 116-120, 204.

32 Theal, *Records of the Cape*, vol. 35, 1897-1905, pp. 94, 98; Yarwood, *Walers*, 1989, pp. 122-123; John Noble (ed.), *Official Handbook of the Cape of Good Hope*, Cape Town: Soloman, 1886, Appendices; *Cape of Good Hope Almanac and Annual Register*, Cape Town, 1845, p. 128; *Official Year Book*, 1922, p. 497; *Parliamentary Papers*, Trade Reports.

33 Rosenthal, *Encyclopaedia*, 1964, p. 235 (and illustration).

34 Marylian and Sanders Watney, *Horse Power*, London: Hamlyn, 1975, pp. 90-91; Wallace, *Farming Industries*, 1896, p. 316.

35 *Official Year Book*, 1922, p. 495; Watney, *Horse Power*, 1975, p. 90; Dirk Postma, *De Trekboeren te St. Januario Humpata*, Amsterdam, 1897, p. 237; William G. Clarence-Smith, *Slaves, Peasants and Capitalists in Southern Angola, 1840-1926*, Cambridge University Press: Cambridge, 1979, pp. 65, 76.

36 British Library, Oriental and India Office Collections, Vanrenen Collection.

37 Tylden, *Horses and Saddlery*, 1980, pp. 28-33, 61-62, 69; Yarwood, *Walers*, pp. 203; Theodore H. Savory, *The Mule, a Historic Hybrid*, Shildon: Meadowfield Press, 1979, p. 23.

38 Richard Pankhurst, *Economic History of Ethiopia, 1800-1935*, Addis Ababa: Haile Selassie I University Press, 1968, pp. 286, 555-556; Callwell, *Small Wars*, pp. 59, 215, 233, 402; Yarwood, *Walers*, pp. 3, 168-170.

39 Tylden, *Horses and Saddlery*, 1980, pp. 62-63.

40 André Schérer, *La Réunion*, Paris: Presses Universitaires de France, 1985, p. 50.

41 Theal, *Records of the Cape*, vol. 32, 1897-1905, 478.

42 Schérer, *La Réunion*, 1985, pp. 81-82; *Encyclopaedia Britannica,* vol. 15, 1929, p. 109.

43 Charles G. Ducray, et al. *Ile Maurice*, Port Louis: General Printing and Stationery Co. Ltd, 1938, pp. 79-83.

44 P. J. Moree, *A Concise History of Dutch Mauritius, 1598-1710*, London: Kegan Paul International, 1998, p. 78.

45 Deryck Scarr, *Seychelles since 1770, History of a Slave and Post-Slavery Society*, London: Hurst & Company, 2000, pp. 12, 17.

46 *The Mauritius Almanac and Colonial Register, 1926-27*, Section A, p. 57; Denis Fielding and Patrick Krause, *Donkeys*, London: Macmillan Education, 1998, p. 90.

47 R. J. Barendse, 'Reflections on the Arabian Seas in the Eighteenth Century', *Itinerario*, vol. 25, no. 1, 2001, p. 30.

48 Robert G. Landen, *Oman since 1856*, Princeton: Princeton University Press, 1967, p. 147; *Mauritius Blue Books*.

49 I. L. Mason and J. P. Maule, *The Indigenous Livestock of Eastern and Southern Africa*, Farnham: Commonwealth Agricultural Bureaux, 1960, p. 16; Epstein, *The Origin*, vol. 2, 1971, pp. 386-387; Patricia W. Romero, *Lamu: History, Society, and Family in an East African Port City*, Princeton: Princeton University Press, 1997, pp. 118, 122-123, 145.

50 *Parliamentary Papers*, Trade Reports for Mauritius; *The Mauritius Almanac and Colonial Register; Mauritius Blue Books*.

51 Pankhurst, *Economic History of Ethiopia*, pp. 211, 352-355, 368, 376, 392, 411, 426, 441-445.

52 *Parliamentary Papers*, Trade Reports for Mauritius.

53 *Mauritius Blue Books*.

54 Blake, *B. I. Centenary*, 1956, pp. 118-119, 127.

55 Ducray, *Ile Maurice*, 1938, pp. 79-81; *Parliamentary Papers*, Trade Reports.

56 Yarwood, *Walers*, 1989, pp. 63-64, 97, 199.

57 Ducray, *Ile Maurice*, 1938, pp. 81-83.

58 I. Gde Parimartha, 'Perdagangan dan Politik di Nusa Tenggara, 1815-1915', Thesis, Vrije Universiteit, Amsterdam, 1995, pp. 104, 208-209; Pieter Hoekstra, *Paardenteelt op het Eiland Soemba*, Batavia, 1948, p. 5.

59 *Mauritius Blue Books*.

60 H. Kistermann, 'De Paardenhandel van Nusa Tenggara, 1815-1941', Student graduation paper, Vrije Universiteit, Amsterdam, 1991, pp. 50, 59.

61 Schérer, *La Réunion*, 1985, pp. 68-69, 80.

62 Peter Boomgaard, 'Horses, Horse Trading and Royal Courts in Indonesian History, 1500-1900'. In Peter Boomgaard and David Henley (eds), *Smallholders and Stockbreeders: Histories of Foodcrop and Livestock Farming in Southeast Asia*, Leiden: KITLV Press, 2004, pp. 211-232.

63 Takeshi Ito, 'The World of the Adat Aceh, a Historical Study of the Sultanate of Aceh', PhD thesis, Australian National University, 1984, pp. 375–379.

64 Serafin D. Quiason, *English 'Country Trade' with the Philippines, 1644–1765,* Quezon City: University of the Pilippines Press, 1966, pp. 17, 73, 94, 105, 172.

65 *British Burma Gazetteer*, 1880, I, p. 435.

66 Max and Bertha Ferrars, *Burma,* London: Sampson Low, Marston & Co., 1900, pp. 139–140.

67 Gervais Courtellemont, *Voyage au Yunnan,* Paris: Plon-Nourrit, 1904, pp. 173–179, 222.

68 William G. Clarence-Smith, 'Horse Breeding in Mainland Southeast Asia and its Borderlands'. In Boomgaard and Henley (eds), *Smallholders and Stockbreeders*, 2004, pp. 189–210.

69 Austin Coates, *China Races*, Hong Kong, 1994.

70 Ernesto J. de Carvalho e Vasconcellos, *As Colónias Portuguesas,* Lisbon, 1896, p. 394; Great Britain, Foreign Office, Historical Section, *Portuguese Timor,* p. 17; Coates, *China Races*, 1994, pp. 3, 15–16, 29, 72.

71 Harry R. Burrill and Raymond F. Crist, *Report on Trade Conditions in China,* New York: Garland Publishing, 1980, p. 68; Yarwood, *Walers,* 1989, pp. 153–156, 168–169, 202; Coates, *China Races*, 1994, pp. 31, 35, 72, 207, 244.

72 Amarjit Kaur, *Bridge and Barrier, Transport and Communications in Colonial Malaya, 1870–1957,* Singapore: Oxford University Press, 1985.

73 *Straits Settlements Annual Reports,* Singapore.

74 Frank J. A. Broeze, 'The Merchant Fleet of Java, 1820–1850: A Preliminary Survey', *Archipel,* vol. 18, 1979, pp. 251–69.

75 J. H. Moor, *Notices of the Indian Archipelago and Adjacent Countries,* London: Oxford University Press, 1968 (reprint of 1837 edn), p. 99; Summerhays, *The Observer's Book,* 1954, p. 65.

76 *Encyclopaedie van Nederlandsch-Indië,* The Hague: W.P. van Stockum & Son, 1917–1921, vol. 3, pp. 151, 227–229; *Straits Settlements Annual Reports.*

77 Peter J. Rimmer, 'Hackney Carriages, Syces and Rikisha Pullers in Singapore: A Colonial Registrar's Perspective on Public Transport, 1892–1923'. In Peter J. Rimmer and Lisa M. Allen (eds), *The Underside of Malaysian History: Pullers, Prostitutes and Plantation Workers,* Singapore: Singapore University Press, 1990, p. 131.

78 Eric Jennings, *The Singapore Turf Club,* Singapore: Singapore Turf Club, 1970.

79 A. Cabaton, *Java, Sumatra and the Other Islands of the Dutch East Indies,* London: T. Fisher Unwin, 1911, pp. 119–120; Robert Kay, 'Java Ponies and Others', *The Horse (Illustrated), The Quarterly Review of the Institute of the Horse and Pony Club,* vol. 11, no. 41, 1939, pp. 24–27.

80 W. B. Worsfold, *A Visit to Java, with an Account of the Founding of Singapore*, London: Bentley and Son, 1893, p. 220; *Encyclopaedie van Nederlandsch-Indië*, vol. 5, p. 59.

81 Gerrit Kuperus, *Het Cultuurlandschap van West-Soembawa*, Groningen: J. B. Wolters, 1936, p. 21.

82 *Koloniaal Verslag 1857*, The Hague, p. 142; *Encyclopaedie van Nederlandsch-Indië*, vol. 3, pp. 151, 227–228.

83 See Peter Boomgaard, Chapter 3, this volume.

84 Broeze, 'The Merchant Fleet of Java', 1979.

85 L.W.C. van den Berg, *Le Hadhramout et les Colonies Arabes dans l'Archipel Indien*, Batavia: Imprimérie du Gouvernement, 1886, p. 150.

86 Kistermann, 'De Paardenhandel', 1991, pp. 51, 62, 65.

87 J. à Campo, *Koninklijke Paketvaart Maatschappij; Stoomvaart en Staatsvorming in de Indonesische Archipel, 1888–1914*, Hilversum: Verloren, 1992.

88 William G. Clarence-Smith, 'Horse Trading: The Economic Role of Arabs in the Lesser Sunda Islands, c. 1800 to c. 1940'. In H. de Jonge and N. Kaptein (eds), *Transcending Borders: Arabs, Politics, Trade, and Islam in Southeast Asia*, Leiden: KITLV Press, 2002, pp. 143–162.

89 *Koloniaal Verslag*, various years; *Encyclopaedie van Nederlandsch-Indië*, vol. 3, p. 151; Parimartha, 'Perdagangan', 1995, pp. 207, 211–212; Kistermann, 'De Paardenhandel', 1991, p. 51; M. H. du Croo, 'Cijfers en Beschouwingen Betreffende het Eiland Soembawa', *Koloniale Studien*, vol. 3, no. 3, 1919, pp. 601–602; Kuperus, *Het Cultuurlandschap*, 1936, p. 188; Hoekstra, *Paardenteelt*, 1948, p. 142; Great Britain, Foreign Office, Historical Section, *Dutch Timor and the Lesser Sunda Islands,* London, 1919, p. 26; Great Britain, Foreign Office, Historical Section, *Portuguese Timor,* 1920, p. 21.

90 Kuperus, *Het Cultuurlandschap*, 1936, p. 27; Parimartha, 'Perdagangan', 1995, pp. 147–148.

91 Berg, *Le Hadhramout*, 1886, p. 146.

92 J. H. P. E. Kniphorst, 'Een Terugblik op Timor en Onderhoorigheden', *Tijdschrift voor Nederlandsch Indië*, vol. 14, no. 2, 1885, pp. 16–17; Kistermann, 'De Paardenhandel', 1991, pp. 38–43.

93 Parimartha, 'Perdagangan', 1995, pp. 160–161.

94 G. C. A. Dijk, 'De Sandelhout en de Paardenfokkerij in Nederlandsch Indië', *De Indische Gids*, vol. 26, no. 1, ii, 1904, pp. 385–386; Kistermann, 'De Paardenhandel', 1991, pp. 55–61; Croo, 'Cijfers', 1919, pp. 603–607; Hoekstra, *Paardenteelt*, p. 46.

95 Yarwood, *Walers*, 1989, pp. 155, 168; *Encyclopaedie van Nederlandsch-Indië*, vol. 3, p. 153.

96 C. A. Heshusius, *KNIL-Cavalerie, 1814-1950; Geschiedenis van de Cavalerie en Pantsertroepen van het Koninklijk Nederlands-Indische Leger*, [no place], c. 1978, pp. 6, 22.

97 C. C. Macknight, 'Outback to Outback: The Indonesian Archipelago and Northern Australia'. In J. J. Fox (ed.), *Indonesia, the Making of a Culture*, Canberra: Australian National University, 1980, p. 144; Yarwood, *Walers*, 1989, p. 165.

98 Yarwood, *Walers*, 1989, pp. 63-64, 149-150, 199.

99 Heshusius, *KNIL-Cavalerie*, 1978, pp. 22-23.

100 Greg Bankoff, 'A Question of Breeding: Zootechny and Colonial Attitudes Towards the Tropical Environment in the Late Nineteenth-century Philippines', *Journal of Asian Studies*, vol. 60, no. 2, 2001, pp. 413-438; Greg Bankoff, 'Horsing Around, The Life and Times of the Horse in the Philippines at the Turn of the Twentieth century'. In Boomgaard and Henley (eds), *Smallholders and Stockbreeders*, 2004, pp. 233-255.

101 John Crawfurd, *A Descriptive Dictionary of the Indian Islands and Adjacent Countries*, Kuala Lumpur: Oxford University Press, 1971, pp. 43, 44, 79, 155; Jean Mallat, *The Philippines: History, Geography, Customs, Agriculture, Industry and Commerce of the Spanish Colonies in Oceania*, Manila: National Historical Institute, 1983, pp. 134, 174.

102 Dan Doeppers, personal communication.

103 John Foreman, *The Philippine Islands,* London: Sampson Low, Marston, Searle & Rivington Ltd, 1890, 1st edn, pp. 301-304.

104 Yarwood, *Walers*, 1989, pp. 150, 165, 199.

105 John Foreman, *The Philippine Islands*, Shanghai, 1906, 3rd edn, pp. 336-338; Hugo H. Miller, *Economics Conditions in the Philippines*, Boston: Ginn & Co., 1920, pp. 326, 330; Charles B. Elliott, *The Philippines to the End of the Commission Government*, New York: Greenwood Press, 1968, p. 346.

106 Yarwood, *Walers*, 1989, pp. 155-156.

Chapter 3. Horse Breeding, Long-distance Horse Trading and Royal Courts in Indonesian History, 1500-1900

Peter Boomgaard

I thank Martine Barwegen, who provided me with various references and with whom I discussed the intricacies of animal husbandry. I am also grateful to David Henley (who also corrected my English) for his comments on an earlier draft. This article was first published – in a slightly different version – in Peter Boomgaard and David Henley (eds), *Smallholders and Stockbreeders: Histories of Foodcrop and Livestock Farming in Southeast Asia*, Leiden: KITLV Press, 2004. It is reprinted by permission of Koninklijk Instituut voor Taal-, Land- en Volkenkunde, Leiden, the Netherlands.

1 Another possibility is that the Portuguese found horses and cattle on these islands when they arrived. The VOC gave these islands Dutch names, such as Delft – the most important 'stud island' – and Twee Gebroeders.

2 Sumbawa and Bima were both 'kingdoms' on the island of Sumbawa, one of the Lesser Sundas (now Nusa Tenggara), as discussed by Bernice de Jong Boers, this volume.

3 *Generale Missiven van Gouverneurs-Generaal en Raden aan Heren XVII der Verenigde Oostindische Companie* [General Missives of the Governors-General and Council to the Gentlemen XVII of the United East Indian Company], edited by W. P. Coolhaas et al, 's-Gravenhage: Nijhoff; Rijks Geschiedkundige Publicatiën, 1960-2004, vol. V, p. 559, vol. VI, p. 169, vol VII, p. 306, 369, vol. VIII, p. 116; F. de Haan, *Priangan: De Preanger-Regentsschappen onder het Nederlandsche Bestuur tot 1811* [Priangan: The Preanger Regencies under Dutch Administration up to 1811], Batavia: Kolff, 4 vols. 1910-12, vol. IV, p. 494; *Selections from the Dutch Records of the Ceylon Government No. 5.* Colombo: State Printing Corporation, 1946, p. 85.

4 *Generale Missiven* vol. VII, p. 686.

5 Jean Gelman Taylor, *Indonesia: Peoples and Histories,* New Haven/London: Yale University Press, 2003, p. 145.

6 O.W. Wolters, *Early Indonesian Commerce: A Study of the Origins of Srivijaya,* Ithaca, New York: Cornell University Press, 1967, p. 59; William G. Clarence-Smith, 'Elephants, Horses, and the Coming of Islam to Northern Sumatra', *Indonesia and the Malay World,* 2004, vol. 32, p. 273.

7 W. P. Groeneveldt, 'Notes on the Malay Archipelago and Malacca, compiled from Chinese Sources', *Verhandelingen Bataviaasch Genootschap,* 1880, vol. 39, pp. 75-98.

8 W. Groeneveldt, 'Het Paard in Nederlandsch-Indië; Hoe het is Ontstaan, Hoe het is en hoe het kan Worden' [The Horse in the Netherlands Indies: How it Originated, How it is, and How it could Become], *Veeartsenijkundige Bladen voor Nederlandsch-Indië,* 1916, pp. 209-211.

9 A. Cortesão (ed.), *The Summa Oriental of Tomé Pires. An Account of the East, from the Red Sea to Japan, Written in Malacca and India in 1512-1515,* London: Hakluyt Society, 1944, vol. I, p. 160.

10 W. P. Groeneveldt, 'Notes on the Malay Archipelago', p. 92.

11 R.C. Temple (ed.) *The Travels of Peter Mundy in Europe and Asia, 1608-1667, vol. III, part 1 (1634-1637),* London: Hakluyt Society (Works issued by the Hakluyt Society, 2nd series, 45), 1919, pp. 121-130; William Dampier, *Voyages and Discoveries* [1699], edited by C. Wilkinson, London: The Argonaut Press 1931, p. 89; Denys Lombard (ed.), *Mémoires d'un Voyage aux Indes Orientales 1619 1622; Augustin de Beaulieu, un Marchand Normand à Sumatra* [no pl.]: Maisonneuve & Larose, 1996, p. 92.

12 The *Daghregister* was the official journal kept by the VOC in Batavia. *Daghregister Gehouden int Casteel Batavia vant Passerende daer ter Plaetse als over Geheel Nederlandts-India* [Daily Register Kept in Castle Batavia of Occurences There and in the Entire Netherlands Indies], 31 vols (covering selected years between 1624 and 1682), Batavia/'s-Gravenhage: Landsdrukkerij/Nijhof, 1644, pp 122, 125; Pieter van Dam, *Beschrijvinge van de Oostindische Compangie* [Descriptions of the East Indian Company], 7 vols, 's-Gravenhage: Martinus Nijhoff: Rijks Geschiedkundige Publicatiën, edited by F.W. Stapel (first 6 vols) and C. W. Th. van Boetzelaer, vol. II-1, p. 261.

13 Crawfurd (1856: 153) recognises at least two races in Sumatra – from Aceh and from Batubara. Groeneveld (1916: 218) calls the Batak horses a 'breed', but he also regards the Batak as one of the four varieties of the Sumatran breed. He recognises a Gayo variety, named after a sparsely populated upland area between Aceh and the Batak region. In a recently published popular booklet on the modern horse breeds of the world, the Batak and the Gayo races are still distinguished as separate breeds, the only Sumatran ones mentioned (Ball 1994: 78).

14 J. Freiherr von Brenner, *Besucht bei den Kannibalen Sumatras. Erste Durchquerung der unabhängigen Batak-Lande* [Visit to the Cannibals of Sumatra. First Cut Right Across the Independent Batak Area] , Würzburg: Woerl, 1894, pp. 343–344; John Anderson, *Acheen, and the Ports on the North and East Coasts of Sumatra*, reprint from the 1st edn 1840, intr. A. J. S. Reid, Kuala Lumpur, etc,: Oxford University Press, 1971, pp. 24, 175, 194, 205; William Marsden, *The History of Sumatra*, reprint of the 3rd edn 1811, intr. by John Bastin, Kuala Lumpur, etc,: Oxford University Press, 1975, pp. 365, 371, 380–382.

15 Stampa [no first name given], 'De paarden in Nederlandsch Oost-Indië' [The Horses in the Netherlands East Indies], *Militaire Spectator; Tijdschrift voor het Nederlandsche Leger,* 1846, vol 14, pp. 172–173; P. Noordijk and J. van der Weijde, 'Staat van den veestapel in de 1e en 3e afdeeling op Java' [Situation of Livestock in the 1st and 3rd Division of Java], *Tijdschrift voor Nijverheid in Nederlandsch Indië,* 1856, vol 3, p. 174; De Haan, *Priangan: De Preanger-Regentschappen,* vol. IV, p. 495.

16 Groeneveldt, 'Notes on the Malay Archipelago', p.50; Peter Boomgaard, *Frontiers of Fear: Tigers, and People in the Malay World, 1600–1950,* New Haven, London: Yale University Press, 2001, pp. 147–148.

17 The Bataks themselves had a 'king', often called the priest-king (Si Singamangaraja). However, this does not seem to have been an ostentatious court such as that of Aceh, so one does not expect a large demand for horses.

18 What I have called the 'Minangkabau breed' is called the 'Sumatran horse' by 't Hoen (1919: 16–17), but given the existence of Batak horses, the term is confusing. Similarly, the Mandailing horse, mentioned by Veth, may be

classified with the Batak breed, a simplification also implicitly offered by 't Hoen. Groeneveld (1916: 216) uses the term 'horse of the Padang Uplands', which is the equivalent of calling it a 'Minangkabau horse'.

19 A. L van Hasselt, *Volksbeschrijving van Midden-Sumatra* [Description of the People of Central Sumatra] (vol. 3 of P. J. Veth (ed.) *Midden-Sumatra: Reizen en Onderzoeking der Sumatra-Expeditie 1877-1879* [Central Sumatra. Travels and Research of the Sumatra Expedition 1877-1879]. Leiden: Brill, 1882, pp. 379-380; Von Brenner, *Besuch bei den Kannibalen*, p. 344; P. J. Veth, *Het Paard onder de Volken van het Maleische Ras* [The Horse among the People of the Malay Race], Leiden: Brill, 1894, p. 29; Christine Dobbin, *Islamic Revivalism in a Changing Peasant Economy: Central Sumatra, 1784-1847*, London, Malmö: Curzon Press, 1983, pp. 15, 66-67, 75.

20 G. P. Rouffaer and J. W. IJzerman (eds), *De Eerste Schipvaart der Nederlanders naar Oost-Indië onder Cornelis de Houtman, 1595-1597; I: D'Eerste boeck, van Willem Lodewyckz* [The First Sea Journey of the Dutch to the East Indies under Cornelis de Houtman, 1595-1597; I: The First Book, by Willem Lodewyckz],'s-Gravenhage: Nijhoff (Werken Linschoten Vereeniging 7) 1915, p. 107; William Foster (ed.), *The Voyage of Sir Henry Middleton to the Moluccas 1604-1606*, London: Hakluyt Society (Works issued by the Hakluyt Society, 2nd series, 88) 1943, p. 161; A. Cortesão (ed.), *The Summa Oriental*, vol I, pp. 167-168.

21 *Daghregister* (1888-1931), 1679: 30 October; *Generale Missiven* (1960-2004) vol. VI, p. 908, vol. VII, p. 577, vol VIII, p. 56; Stampa, 'De paarden', 1846, p. 171; J. K. J de Jonge (& M. L. van Deventer) *De Opkomst van het Nederlandsch Gezag in Oost-Indië. Versameling van Onuitgegeven Stukken uit het Oud-koloniaal Archief (1595-1814)* [The Rise of Dutch Power in the East Indies. Collection of Unpublished Documents from the Old-Colonial Archives (1595-1814)], 17 vols, 's-Gravenhage: Nijhoff, 1862-95, vol V, p. 241; P. J. Veth, *Het Paard*, pp. 57-58.

22 *Generale Missiven* (1960-2004), vol. VI, pp. 767, 890, vol. VII, pp. 196, 761, vol. VIII, pp. 26-7, 93, vol. IX, p. 60; De Haan, *Priangan: De Preanger-Regentschappen*, vol. IV, p. 495.

23 *Daghregister* (1888-1931), 1680: 2 August; Stampa, 'De paarden', 1846, pp. 171-172; P. Noordijk and J. van der Weijde, 'Staat van den veestapel', pp. 178-179; J. K. J. de Jonge (& M. L. van Deventer) *De Opkomst van het Nederlandsch Gezag*, vol V, p. 94; P. J. Veth, *Het Paard*, pp. 30, 32; De Haan, *Priangan: De Preanger-Regentschappen*, vol. IV, p. 499.

24 P. P. Roorda van Eysinga, *Verschillende Reizen en Lotgevallen van S. Roorda van Eysinga* [Various Travels and Adventures of S. Roorda van Eysinga]. 4 vols, Amsterdam, van der Heij (1830-2), vol III, p. 66; Stampa, 'De paarden', 1846, p. 171; P. Noordijk and J. van der Weijde, 'Staat van den veestapel', pp. 174-176; P. J. Veth, *Het Paard*, pp. 30, 32; De Haan, *Priangan: De Preanger-*

Regentschappen, vol. IV, pp. 493-499; H.'t Hoen, *Veerassen en Veeteelt in Nederlandsch-Indië* [Livestock Races and Stockbreeding in the Netherlands Indies], Weltevreden, Kolff, 1919, pp. 9-10.

25 Stampa, 'De paarden', 1846, p. 171; P. Noordijk and J. van der Weijde, 'Staat van den veestapel', pp. 174-175.

26 Nationaal Archief (The National Archives in The Hague (hereafter NA)), Ministerie van Koophandel en Koloniën, 147: General Inspection Report, Priangan, 29 January 1808, par. 202

27 P. Noordijk and J. van der Weijde, 'Staat van den veestapel', pp. 174-176; W. Groeneveld, 'Het Paard', p. 218.

28 *Daghregister* (1888-1931), 1625: 20 May, 1631: 22 July; J. K. J. de Jonge (& M. L. van Deventer) *De Opkomst van het Nederlandsch Gezag*, vol IV, p. 220, vol. V, pp. 139, 197-198; Pieter van Dam, *Beschrijvinge van de Oostindische*, vol. II-3, pp. 448-449.

29 J. K. J. de Jonge (& M. L. van Deventer) *De Opkomst van het Nederlandsch Gezag*, vol. VII, pp. 145, 216, 236, 242; Pieter van Dam, *Beschrijvinge van de Oostindische*, vol. II-3, p. 403.

30 NA, Collection Reinwardt, 17: Fischer, *Beschrijvinge*, c. 1780; Stampa, 'De paarden', 1846, pp. 172-173; *De Residentie Kadoe naar de Uitkomsten der Statistieke Opname* [The Residency of Kadoe Based on the Results of the Statistical Survey], Batavia: Landsdrukkerij, 1871, p 118; P. J. Veth, *Het Paard*, pp. 32-38; H.'t Hoen, *Veerassen en Veeteelt*, pp. 10, 14-15.

31 Franz Wilhelm Junghuhn, *Java, zijne Gedaante, zijn Plantetooi en Inwendige Bouw* [Java, its Appearance, its Flora, and its Morphology], 3 vols, 's-Gravenhage: Mieling, 1853-54; Peter Boomgaard, 'Maize and Tobacco in Upland Indonesia'. In Tania Murray Li (ed.), *Transforming the Indonesian Uplands: Marginality, Power and Production*, Amsterdam: Harwood Academic Publishers, 1999, pp. 53-60.

32 A. Cortesão (ed.), *The Summa Oriental*, vol I, pp. 174-176, 191.

33 NA, Collection Reinwardt, 17: Fischer, *Beschrijvinge*, c. 1780; Stampa, 'De paarden', 1846, p. 171; J. van der Weide, 'Iets over de op Java voorkomende paarden', *Tijdschrift voor Nijverheid in Nederlandsch-Indië*, 1860, pp. 390; P. H. van der Kemp, 'Historisch overzicht van de pogingen aangewend tot verbetering en veredeling van het paardenras in Nederlandsch-Indië' [Historical Overview of the Attempts to Improve the Breed of Horses in the Netherlands Indies], *Veeartsenijkundige Bladen voor Nerderlandsch-Indië*,1890, vol. 4, p. 363; W. Groeneveld, 'Het Paard', p. 221; H. 't Hoen, *Veerassen en Veeteelt*, p. 9; A. Cortesão (ed.), *The Summa Oriental*, vol. I, pp. 197-198, 227.

34 A. Cortesão (ed.), *The Summa Oriental*, vol. I, pp. 202-203, P. J. Veth, *Het Paard*, pp. 50-53; H. 't Hoen, *Veerassen en Veeteelt*, pp. 17-22.

35 W. Groeneveld, 'Het Paard', p. 50; T. G. T. Pigeaud, *Java in the 14th Century; a Study in Cultural History,* 5 vols, The Hague: Nijhoff (KITLV Translation series, 4), 1960-63, vol. IV, p. 519; A. M. Barrett Jones, *Early Tenth Century Java from Inscriptions,* Dordrecht, Cinnaminson: Foris (Verhandelingen KITLV, 107), pp. 57-58.

36 Figures for 1820/2, by Residency, are to be found in NA, Collection Schneither, 83-100: Statistical Accounts of the Java Residencies, c. 1820. On India c. 1600, see Foster 1921: 17, 103-104.

37 NA, Collection Van Alphen/Engelhard, 1900, 191: Memoir of Succession, Java's Northeast Coast, 1801, par. 393; Idem, 1916, 104: Report of the Regent of Batang, c. 1806, article 23.

38 M. D. Teenstra, *De Vruchten mijner Werkzaamheden, Gedurende mijne Reize naar Java* [The Fruits of my Labours, During my Journey to Java], 3 vols, Groningen: Eekhoff, 1828-30, vol. II, pp. 97-98; P. P. Roorda van Eysinga, *Verschillende Reizen,* vol. I, p. 197, vol. II, p. 369; De Haan, *Priangan: De Preanger-Regentschappen,* vol. IV, pp. 487, 493-499.

39 *Generale Missiven* (1960-2004), vol. V, pp. 126, 754, vol. VII, pp. 97, 234, 299, 359, 562, 617; De Haan, *Priangan: De Preanger-Regentschappen,* vol. IV, p. 495.

40 These prices probably included the cost of transportation to Siam and an interest payment on the loan.

41 John Crawfurd, *A Descriptive Dictionary of the Indian Islands & Adjacent Countries,* London: Bradbury & Evans, 1956, pp. 153-155; E. Francis, *Herinneringen uit den Levensloop van een 'Indisch' Ambtenaar van 1815 tot 1851* [Memoirs from the Career of an 'Indian' Civil Servant from 1815 to 1851], Batavia, Van Dorp, 3 vols. 1856-9, vol. II, pp. 135, 161-163; W. J. L. Wharton (ed.), *Captain Cook's Journal during his First Voyage round the World made in H.M. Bark ' Endeavour' 1768-71,* London: Stock, 1893, p. 344; J. Veth, *Het Paard,* pp. 39-50; H. 't Hoen, *Veerassen en Veeteelt,* pp. 17-22; William Dampier, *A Voyage to New Holland,* edited by J. A. Williamson, London: Argonaut Press, 1939, pp. 165-170; A. Cortesão (ed.), *The Summa Oriental,* vol. I, pp. 202-203; James J. Fox, *Harvest of the Palm: Ecological Changes in Eastern Indonesia,* Cambridge, London: Harvard University Press, 1977, pp. 21, 162; Bernice de Jong Boers, 'Paardenfokkerij op Sumbawa (1500-1930)' [Horse Breeding in Sumbawa (1500-1930)], *Spiegel Historiael,* vol. 32. no. 10/11, pp. 438-443; William Gervase Clarence-Smith, 'Horse Trading: The economic role of Arabs in the Lesser Sunda Islands, c. 1800 to c. 1940'. In Huub de Jonge and Nico Kaptein (eds), *Transcending Borders: Arabs, Politics, Trade and Islam in Southeast Asia,* Leiden: KITLV Press, 2002, pp. 143-162.

42 *Daghregister* (1888-1931), 1672: 27 July, 1673: 22 August, 1680: 19 December, 1682: 10 March; Raffles 1830, vol. I, pp. 53-4; E. Francis, *Herinneringen,* vol. III, p. 135; A. Cortesão (ed.), *The Summa Oriental,* vol. I, pp. 202-203.

43 W. van Hogendorp, 'Beschryving van het Eiland Timor, voor Zoo Verre het tot Nog toe Bekind is', *Verhandeling Bataviaasch Genootschap,* vol. 2, p. 81; E. Francis, *Herinneringen,* vol. II, pp. 161-163; Wharton (ed.), *Captain Cook's Journal,* p. 344; William Dampier, *A Voyage to New Holland,* 1939, p. 168.

44 This statement is deduced from information on the import of Bima horses in Makassar at an early stage.

45 Valentijn 1724/6, vol. III-2, p. 138; John Crawfurd, *A Descriptive Dictionary,* pp. 89-90; Von Brenner, *Besuch bei den Kannibalen,* p. 344; J. Veth, *Het Paard,* pp. 25, 51-53; De Haan, *Priangan: De Preanger-Regentschappen,* vol. IV, p. 494; H. 't Hoen, *Veerassen en Veeteelt,* p. 24.

Chapter 4. The 'Arab' of the Indonesian Archipelago: Famed Horse Breeds of Sumbawa
Bernice de Jong Boers

This chapter is partly based on two unpublished papers and one published article, namely: Bernice de Jong Boers, 'Horses: One of the First Commodities of the Island of Sumbawa (Indonesia)', Gadjah Mada University and Leiden University, paper written for the first Summer Course in Indonesian Modern Economic History, 1995; Bernice de Jong Boers, 'Some Notes on the History of Livestock in Indonesia (with a special focus on Sumbawa)', paper presented at Animals in Asia: Relationships and Representations, 15-16 September 1997; and Bernice de Jong Boers, 'Paardenfokkerij op Sumbawa (1500-1930)' [Horse Breeding on the Island of Sumbawa (1500-1930)], *Spiegel Historiael,* vol. 32, no. 10/11, 1997, pp. 438-443, supplemented with new data from historical sources. I would like to thank Greg Bankoff and Sandra Swart for their useful comments on an earlier version of this article, and Youetta de Jager, Leendert Bergman and André van der Ham for assistance with the production of the maps.

Topographical names
In the text various spellings and names are used for the same topographical places, regions and islands. Sometimes the current (Indonesian) spelling and name is used and at other times the (Dutch) colonial spelling or name, depending on the context. What follows is a short list of these place names beginning first with their current spelling followed between brackets by the old or colonial usage: Bantaeng (Bonthain), Gowa (Goa), Jakarta (Batavia), Manado (Menado), Malacca (Malakka), Reunion (Bourbon), Riau (Riouw), Sri Lanka (Ceylon), Sulawesi (Celebes), Sumbawa (Soembawa). Borneo is the name of the whole island while Kalimantan refers only to the Indonesian part of it. The Lesser Sunda Islands comprise the islands of Bali, Lombok, Sumbawa, Komodo, Flores, Timor, Sawu and Roti. Today, all these islands except Bali constitute the region of Nusa Tenggara. Finally, Laambu, Kangga, Paie, Poja, Wera, and

Saie as mentioned in the text are all villages or regions in the province of Bima (Sumbawa).

1 John Crawfurd, *A Descriptive Dictionary of the Indian Islands and Adjacent Countries*, London: Bradbury and Evans, 1856, p. 153.

2 See also, Peter Boomgaard, 'Horses, Horse Trading and Royal Courts in Indonesian history, 1500-1900'. In Peter Boomgaard and David Henley (eds), *Smallholders and Stockbreeders: History of Foodcrop and Livestock Farming in Southeast Asia*, Leiden: KITLV Press, 2004, pp. 211-232.

3 *Ladang* is a dry rice field or swidden.

4 According to the climate classification developed by the Russian/Austrian biologist and climatologist Vladimir Köppen. See: www.britannica.com/eb/article?tocId=9046045 (23.1.2005) and www.britannica.com/eb/article?tocId=53348 (23.1.2005).

5 These princedoms were also called sultanates after the introduction of Islam in Sumbawa at the beginning of the seventeenth century. To avoid confusion the princedom of Sumbawa will be distinguished from the island as a whole by using the names 'Western Sumbawa' when the region or princedom is meant and 'Sumbawa' when the island is meant.

6 W. Groeneveld, and J. C. Witjens, *Het Paard* [The Horse], Haarlem: Tjeenk Willink, Onze Koloniale Dierenteelt, *Populaire Handboekjes over Nederlandsch-Indische Nuttige Dieren*, 1924, p. 8.

7 Ibid., p. 3.

8 Ibid., pp. 1-2.

9 Jan Wisseman Christie, 'The Agricultural Economies of Early Java and Bali'. In Peter Boomgaard and David Henley (eds), *Smallholders and Stockbreeders*, 2004, pp. 47-68; and Peter Boomgaard, Chapter 3, this volume.

10 Gerrit Kuperus, Het Cultuurlandschap van West-Soembawa [The Man-Made Landscape of Western-Sumbawa], Groningen, Batavia: J. B. Wolters, 1936, p. 133; and Peter Goethals, *Aspects of Local Government in a Sumbawan Village (Eastern Indonesia)*, Ithaca, New York: Cornell University, Monograph Series, Modern Indonesia Project, Southeast Asia Program, Department of Far Eastern Studies, 1961, p. 9.

11 Nicolaus Adriani, *Bare'e-Nederlandsch Woordenboek* [Bare'e-Dutch Dictionary], Leiden: Brill, Bataviaasch Genootschap van Kunsten en Wetenschappen, 1928.

12 William Thorn, *Memoir of the Conquest of Java with the Subsequent Operations of the British Forces in the Oriental Archipelago*, Singapore: Periplus Editions, 1993, p. 319. Thorn (1781-1843), a British Major who joined the British Forces serving in Java, also sailed to other regions in the archipelago.

His memoir about the military operations and journeys in the archipelago was first published in 1815.

13 Crawfurd, *A Descriptive Dictionary of the Indian Islands*, 1856, p. 153. A Scotsman, John Crawfurd, was trained as a doctor and was employed by the British East Asia Company. He worked in India, Malaysia and Indonesia. When Thomas Stamford Raffles was appointed Lieutenant-Governor in 1811, Crawfurd was appointed as the Resident of Yogyakarta (Java) and was sent on several diplomatic missions throughout the archipelago.

14 W. H. Davenport Adams, *The Eastern Archipelago*, London: T. Nelson and Sons, 1880, p. 199; and G. W. Couperus, 'Les Races chevalines des Iles de la Sonde', *Revue Coloniale Internationale*, vol. 2, 1886, p. 31.

15 Johannes Olivier, *Land- en Zeetogten in Nederlands Indië, en Eenige Britsche Etablissementen, Gedaan in de Jaren 1817 tot 1826* [Journeys by Land and Sea in the Netherlands-Indies, and Some British Establishments, Made in the Years Between 1817 and 1826], Part II, Amsterdam: Sulpke, 1828, p. 240. Olivier (1790–1858) was a Dutch official who stayed in and travelled to several places in the archipelago working as a writer and as a teacher.

16 H. 't Hoen, *Veerassen en Veeteelt in Nederlandsch-Indië* [Cattlebreeds and Cattle Breeding in the Netherlands-Indies], Batavia: Kolff, 1919, pp. 17–19.

17 Bernice de Jong Boers, 'Tambora 1815: De Geschiedenis van een Vulkaanuitbarsting in Indonesië' [Tambora 1815: The History of a Volcanic Eruption in Indonesia], *Tijdschrift voor Geschiedenis*, vol. 107, no. 3, 1994, pp. 371–392; and Bernice de Jong Boers, 'Mount Tambora in 1815: A Volcanic Eruption in Indonesia and its Aftermath', *Indonesia*, vol. 60, 1995, pp. 36–60.

18 *Singapore Chronicle* 1825; Couperus, 'Les Races Chevalines des Iles de la Sonde', 1886, p. 31; Pieter Johannes Veth, *Het Paard Onder de Volken van het Maleische Ras* [The Horse among the People of the Malay Race], Leiden: E. J. Brill, 1894, p. 39; and Daphne Machin Goodall, *Paarderassen van de Wereld* [Horses of the World], Zwolle: La Rivière en Voorhoeve, 1968, p. 176.

19 F. H. H. Guillemard, The Cruise of Marchesa to Kamschatka and New Guinea (with Notices of Formosa, Liu-Kiu, and Various Islands of the Malay Archipelago), London: John Murray, 2nd edn, 1889, p. 274; and Heinrich Zollinger, 'Verslag van eene Reis naar Bima en Soembawa en naar Eenige Plaatsen op Celebes, Saleyer, en Floris Gedurende de Maanden Mei tot December 1847' [Report of a Journey to Bima and Sumbawa and to Several Places on Celebes, Saleyer and Flores during the Months May to December 1847], Verhandelingen van het Bataviaasch Genootschap voor Kunsten en Wetenschappen (VBG), vol. 23, 1850, p. 78. The original text in Dutch states: 'Het voornaamste huisdier van het eiland zal wel het paard zijn, dat nergens zooveel wordt aangetroffen als op Bima en Soembawa. Dat het ras een der beste is van den gehelen

Archipel, is bekend. Ja, velen houden het voor het beste.' Heinrich Zollinger (1818-1859) was a naturalist from Zürich, Switzerland. He undertook several research expeditions in the archipelago and was dispatched to the island of Sumbawa in 1847 by the then Governor-General of the Netherlands East Indies, Jan Jacob Rochussen (De Jong Boers 2000/2001: 12). Bernice de Jong Boers, 'Een Proefschrift in Wording over de Milieugeschiedenis van het Eiland Sumbawa, Indonesië' [A Note on Writing a PhD thesis about the Environmental History of the Island of Sumbawa, Indonesia], De Boekerij: Mededelingenblad van de Vereniging Vrienden der Bibliotheek van de Koninklijke Nederlandse Akademie van Wetenschappen, vol. 5, no. 3/vol.6, no. 1, p. 12.

20 *Singapore Chronicle* 1825; and Crawfurd, *A Descriptive Dictionary of the Indian Islands and Adjacent Countries*, 1856, p. 153. Isabel is a term used in equine literature to describe a particular skin colour of horses. It is a yellowish colour against a light or dark skin, sometimes also called palomino.

21 Armando Cortesão (ed.), The Suma Oriental of Tomé Pires: An Account of the East, from the Red Sea to Japan, Written in Malacca and in India 1512-1515, and the Book of Francisco Rodrigues Rutter of a Voyage in the Red Sea, Nautical Rules, Almanack and Maps, Written and Drawn in the East before 1515, London: The Hakluyt Society (second series, no. XC and LXXXIX), 1944, p. 203.

22 Mansel Longworth Dames (ed.), The Book of Duarte Barbosa: An Account of the Countries Bordering on the Indian Ocean and their Inhabitants, Written by Duarte Barbosa and Completed about the Year 1518 A.D., London: Hakluyt Society, vol. 49, 1918, p. 194.

23 Zollinger, 'Verslag van eene Reis naar Bima en Soembawa', 1850, p. 106.

24 Henri Chambert-Loir, 'State, City, Commerce: The Case of Bima', *Indonesia*, vol. 57, 1994, p. 80.

25 Ibid., p. 133. The *Jene Jara Asi* and *Jene Jara Kapa* were the commanders of the horsemen and troopers.

26 D. F. van Braam Morris, 'Nota van Toelichting bij het Contract Gesloten met het Landschap Bima 1886' [Explanatory Memorandum to the Contract concluded with the Region of Bima 1886], *Tijdschrift voor Indische Taal-, Land- en Volkenkunde (TBG)*, vol. 34, 1891, pp. 202-203. D. F. van Braam Morris was Governor of Celebes and Dependencies in 1888.

27 M. A. Bouman, 'Toeharlanti: De Bimaneesche Sultansverheffing' [Tuharlanti: the Bimanese Installation of a Sultan], *Koloniaal Tijdschrift*, vol. 14, 1925, p. 713. I am not quite sure whether Bouman's translation is correct. In Indonesian *Jara Manggila* would mean something like 'the horse that ran wild'. In the Manggarai language, a language spoken in West-Flores near Bima, *Menggila* refers to either a fairy-tale horse or to feral horses on the island of Rinca. Jilis Verheijen, *Kamus Manggarai I: Manggarai-Indonesia* [A

Dictionary of Manggarai I: Manggarai-Indonesian], 's-Gravenhage: Martinus Nijhoff (KITLV), 1967, p. 324.

28 Van Braam Morris, 'Nota van Toelichting bij het Contract Gesloten met het Landschap Bima 1886', 1891, p. 216; H. St. Maryam R. Salahuddin and H. Abdul Wahab H. Ismail, *Pemerintah Adat Kerajaan Bima: Struktur dan Hukum* [Adat Prescriptions of the Kingdom of Bima: Structure and Legislation], Mataram: Departemen Pendidikan dan Kebudayaan [Museum Negeri Nusa Tenggara Barat], 1988/1989, pp. 22-23.

29 W. G. C. toe Water, 'Sampela, een Tafereel van Bimanesche Zeden, Gewoonten en Karakters' [Sampela, a Scene of Bimanese Manners, Customs and Characters], *Tijdschrift van Nederlandsch-Indië (TNI),* vol. 6, 1844, p. 552; Van Braam Morris, 'Nota van Toelichting bij het Contract Gesloten met het Landschap Bima 1886', 1891, p. 215; and J. E. Jasper, 'Het Eiland Soembawa en zijn Bevolking' [The Island of Sumbawa and its Population], *Tijdschrift voor het Binnenlandsch Bestuur,* vol. 34, 1908, p. 95.

30 Van Braam Morris, 'Nota van Toelichting bij het Contract Gesloten met het Landschap Bima 1886', 1891, p. 215; and Jasper, 'Het Eiland Soembawa en zijn Bevolking', 1908, p. 95. *Reals* were coins made of gold, silver or copper used in Dutch districts between the sixteenth and eighteenth centuries under Spanish–Portuguese influence.

31 Jasper, 'Het Eiland Soembawa en zijn Bevolking', 1908, p. 95; he was a Dutch official who visited Sumbawa around 1908.

32 Bouman, 'Toeharlanti: De Bimaneesche Sultansverheffing', 1925, pp. 712-715.

33 M. Syaraswati and Yusuf H. Umar, *Upacara U'a Pua di Kabupaten Bima (Pengaruh Agama Islam)* [The U'a Pua Ceremony in Kabupaten Bima (the Influence of the Religion of Islam)], Mataram: Departemen Pendidikan dan Kebudayaan, Museum Negeri Nusa Tenggara Barat, 1985/1986, pp. 29-30.

34 Thorn, *Memoir of the Conquest of Java,* 1993 [1815], p. 319.

35 A story told to Michael Hitchcock by members of the Bimanese cultural organisation '*La Mbila*', Michael J. Hitchcock, 'Technology and Society in Bima, Sumbawa, with Special Reference to House Building and Textile Manufacture', PhD thesis, Department of Ethnology and Prehistory, University of Oxford, 1983, p. 39. Prince Diponegoro (1785-1855) from Yogyakarta, son of Sultan Hamengkubuwana III, unleashed a revolt against the Dutch after a dispute over a new road near the village of Tegalreja which eventually resulted in the bloody Java War between 1825 and 1830. Dutch authority in Central and East Java was shaken but after the rebels had been severely weakened both by disease (cholera, malaria and dysentery) and by the surrender of his uncle, Diponegoro entered into negotiations with the Dutch in 1830. The rebellion ended and Diponegoro was exiled to Menado and Makassar. Around 200,000 Javanese perished during the war as well as

8,000 European and 7,000 Indonesian soldiers. Merle Ricklefs, *A History of Modern Indonesia*, London: Macmillan (Macmillan Asian Histories Series), 1981, pp. 111–113. Today, Diponegoro is one of Indonesia's national heroes and remembered as an early freedom fighter against the Dutch.

36 Zollinger, 'Verslag van eene Reis naar Bima en Soembawa', 1850, p. 105. The *Ruma Bicara* (*Raja Bicara* in Indonesian) was a functionary directly under the Sultan and translated in Western sources as Chief Minister. Chambert-Loir, 'State, City, Commerce', 1994, p. 97.

37 Van Braam Morris, 'Nota van Toelichting bij het Contract Gesloten met het Landschap Bima 1886', 1891, p. 190 and 222; and J. P. Freijss, 'Schets van den Handel van Sumbawa' [Sketch of the Trade on Sumbawa], *Tijdschrift van Nederlandsch-Indië* (TNI), vol. II, 1859, p. 272.

38 Albertus Ligtvoet, 'Aanteekeningen Betreffende den Economischen Toestand en de Ethnographie van het Rijk van Soembawa' [Notes Concerning the Economic Situation and the Ethnography of the Realm of Sumbawa], *Tijdschrift voor Indische Taal-, Land- en Volkenkunde (TBG)*, vol. 23, 1876, pp. 555–570.

39 Ligtvoet, 'Aanteekeningen Betreffende den Economischen Toestand en de Ethnographie van het Rijk van Soembawa', 1876, p. 567 and 589–590. Ligtvoet was a Dutch official stationed in Makassar at the service of the Resident of Celebes and Dependencies between 1864 and 1879. J. Noorduyn, *Bima en Sumbawa: Bijdragen tot de Geschiedenis van de Sultanaten Bima en Sumbawa door A. Ligtvoet en G .P. Rouffaer* [Bima and Sumbawa: Contributions to the History of the Sultanates of Bima and Sumbawa by A. Ligtvoet and G. P. Rouffaer], Dordrecht: Foris (KITLV, Verhandelingen 129), 1987, p. 1. At that time, the island of Sumbawa resided under this Residency.

40 Jasper, 'Het Eiland Soembawa en zijn Bevolking', 1908, p. 94.

41 *Cidomo, dokar* and *Ben Hur* are all local names for small carts and coaches pulled by a horse.

42 MMK 345, Memorie van Overgave [Memorandum of Transfer] (hereafter MvO), J. J. Bosch, Timor en Onderhorigheden [Timor and Dependencies], 29.3.1938, p. 99.

43 Gerrit Kuperus, 'Beschouwingen over de Ontwikkeling en den Huidigen Vormenrijkdom van het Cultuurlandschap in de Onderafdeeling Bima (Oost-Soembawa)' [Considerations on the Genesis of the Current Polymorphology in the Man-Made Landscape in the Bima District (Eastern Sumbawa)], *Tijdschrift van het Koninklijk Aardrijkskundig Genootschap (KNAG)*, vol. 55, no. 2, 1938, p.13. Kuperus was a scholar who took his doctoral degree in 1936 based on a geographical study of Western Sumbawa. He visited the island in 1931 and 1932.

44 Van Braam Morris, 'Nota van Toelichting bij het Contract Gesloten met het Landschap Bima 1886', 1891, p. 190; Michael Dove, 'Man, Land and Game in

Sumbawa, Eastern Indonesia', *Singapore Journal of Tropical Agriculture,* vol. 5, no. 2, 1984, p. 116.

45 H. Kistermann, 'De Paardenhandel van Nusa Tenggara 1815-1941' [The Horse Trade in Nusa Tenggara], Amsterdam: Vrije Universiteit, unpublished doctoral thesis, 1990, pp. 26-27.

46 Overgekomen Brieven en Papieren [Transmitted Letters and Papers] (hereafter OBP), The Hague: National Archive, VOC archives, OBP 1240 and 1246. The trade was especially in sappan wood – a product of the wood from which red paint pigment (dyestuff) was prepared.

47 KIT 1213, Militaire Memorie over de Afdeeling Soembawa [Military Report on the Region of Sumbawa], anonymous, 1929.

48 OBP 1287, folio 1232-1233.

49 Boomgaard, Chapter 3, this volume.

50 Jacob Cornelis Matthieu Radermacher, 'Korte Beschrijving van het Eiland Celebes en de Eilanden Floris, Sumbauwa, Lombok en Bali' [Short Description of the Island of Celebes and the Islands Flores, Sumbawa, Lombok and Bali], *Verhandelingen van het Bataviaasch Genootschap voor Kunsten en Wetenschappen (VBG),* vol. 4, 1786, p. 186 and Thorn, *Memoir of the Conquest of Java,* 1993 [1815], p. 320.

51 De Jong Boers, 'Mount Tambora in 1815', 1995.

52 Caspar Georg Carl Reinwardt, *Reis naar het Oostelijk Gedeelte van den Indischen Archipel, in het Jaar 1821* [Journey to the Eastern Part of the Indian Archipelago, in the Year 1821], edited by W. H. de Vriese, Amsterdam: Frederik Muller, 1858, p. 317.

53 Ibid., p. 318. C. G. C. Reinwardt (1773-1854) was a Prussian-born Dutch botanist who established the ' 's-lands Plantentuin' at Bogor (the Bogor Botanical Garden) nowadays called *Kebun Raya* in 1817. He spent six years in the Indonesian archipelago. After his return to the Netherlands in 1822, he took up a professorship at Leiden University in 1823.

54 Emanuel Francis, *Herinneringen uit de Levensloop van een Indisch Ambtenaar van 1815 tot 1851* [Memories from the Course of the Life of an 'Indisch' Official from 1815 to 1851], 3 vols, Batavia: Van Dorp, 1856, p.134. Francis was a Dutch official who was stationed in various regions of the Netherlands East Indies between 1815 and 1851.

55 Zollinger, 'Verslag van eene Reis naar Bima en Soembawa', 1850, pp. 105, 107-108 and 114.

56 Crawfurd, *A Descriptive Dictionary of the Indian Islands,* 1856, p. 153; Guillemard, *The Cruise of Marchesa to Kamschatka and New Guinea,* 1889, p. 153; and J. Ballot, 'Historisch Overzicht van de Maatregelen tot Verbetering van den Paarden- en Veestapel in Nederlandsch-Indië' [Historical Review of

Measures to Improve the Horse and Cattle Population in the Netherlands-Indies], *Veeartsenijkundige Bladen voor Nederlandsch-Indië*, vol. 10, 1897, p. 28.

57 Freijss, 'Schets van den Handel van Sumbawa', 1859, p. 272. Freijss (his name is also spelled as Freijs and Freys) was a merchant in Lombok and often paid visits to Sumbawa for business purposes.

58 *Jaarverslag van de Burgerlijke Veeartsenijkundige Dienst* (BVD) [Yearly report of the Civil Veterinary Service], 29 vols, covering selected years between 1922 and 1940, Batavia: Departement van Economische Zaken, 1923–1941, see years 1922–1938; and *Jaarboek van het Departement van Landbouw, Nijverheid en Handel in Nederlandsch-Indië* [Year-book of the Department of Agriculture in the Netherlands-Indies], 24 vols, covering selected years between 1906 and 1929, Batavia: Landsdrukkerij; Weltevreden: Albrecht, 1906-1911 and 1914-1921.

59 Jaarverslagen van de BVD, 1922–38. *Gerobak* means cart or coach.

60 MvO, J. J. Bosch, 29.3.1938, line 23; and Michael Georg de Boer and Johannes Cornelis Westermann, *Een Halve Eeuw Paketvaart, 1891-1941* [Half a Century of Packet-Shipping], Amsterdam: DeBussy, 1941, pp. 276–279.

61 A *kuli* is a day-labourer mostly hired for tough jobs, in English called coolie.

62 De Boer and Westermann, *Een Halve Eeuw Paketvaart*, 1941, pp 281–282.

63 Maurice Henri du Croo, 'De Verbetering van den Paardenstapel op het Eiland Soembawa' [The Improvement of the Horse Population on the Island of Sumbawa], *Koloniale Studiën*, vol. 1 (Part II), 1917, p. 479 and J. Sibinga Mulder, 'De Economische Beteekenis van het Vee in Nederlandsch Oost-Indië en de Regeeringszorg Ervoor' [The Economic Significance of Livestock in the Netherlands East Indies and the Government Measures], *De Indische Gids*, vol. 19 (Part I), 1927, p. 323.

64 These measures were taken within the broader context of the so-called 'Ethische Politiek' (Ethical Policy) which was officially launched in 1901. This policy was a kind of welfare programme, called into being because the Dutch felt the moral obligation to compensate for 'centuries of exploitation'. Robert Cribb (1993) 'Development Policy in the Early 20th Century'. In J. P. Dirkse, Frans Hüsken and Mario Rutten, *Development and Social Welfare Indonesia's Experiences under the New Order*, Leiden: KITLV Press, 1993, p. 226.

65 Du Croo, 'De Verbetering van den Paardenstapel op het Eiland Soembawa', 1917, p. 481; and Sibinga Mulder, 'De Economische Beteekenis van het Vee in Nederlandsch Oost-Indië en de Regeeringszorg Ervoor', 1927, pp. 323–324.

66 Du Croo, 'De Verbetering van den Paardenstapel op het Eiland Soembawa', 1917; and Ajoebar, 'De Verbetering van den Paardenstapel op West-Soembawa' [The Improvement of the Horse-Population in Western-Sumbawa], *Koloniale*

Studiën, vol. 11 (Part II), 1927, pp. 366–373. See also Greg Bankoff, Chapter 6, this volume.

67 Jaarverslagen van de BVD, 1922–38.

68 MMK 342, MvO, C. Schultz, Timor en Onderhorigheden [Timor and Dependencies], 13.6.1927, pp. 8 and 37; MMK 343, MvO, P. F. J. Karthaus, Timor en Onderhorigheden [Timor and Dependencies], 6.5.1931, p. 22; and MMK 344, MvO, E. H. de Nijs Bik, Timor en Onderhorigheden [Timor and Dependencies], 16.6.1934, p. 61.

69 Raden Soekardjo Sastrodihardjo, *Beberapa Tjatatan Tentang Daerah Pulau Sumbawa* [Some Remarks Concerning the Island of Sumbawa], Singaradja: Djawatan Petanian Rakjat Propinsi Nusa Tenggara, 1956, pp. 43–44; *Nusa Tenggara dalam Angka, 1978* [Nusa Tenggara in Statistics, 1978], Mataram: Kantor Sensus dan Statistik Propinsi NTB, 1979, p. 136; and *Nusa Tenggara dalam Angka, 1987* [Nusa Tenggara in Statistics, 1987], Mataram: BAPPEDA dan Kantor statistik propinsi NTB, 1988, p. 133.

70 Sibinga Mulder, 'De Economische Beteekenis van het Vee in Nederlandsch Oost-Indië en de Regeeringszorg Ervoor', 1927, pp. 311.

71 Zollinger, 'Verslag van eene Reis naar Bima en Soembawa', 1850, p. 176; and Francis, *Herinneringen uit de Levensloop van een Indisch Ambtenaar*, 1856, p. 131.

72 *Sawah* is a wet rice field (sometimes dependent on rain, sometimes artificially irrigated), in English called a paddy field.

73 't Hoen, *Veerassen en Veeteelt in Nederlandsch-Indië*, 1919, p. 20; and 'Het Veeteeltbedrijf in den Ambtskring Soembawa-Besar' [Animal Husbandry in the District Sumbawa-Besar], *Nederlands-Indische Bladen voor Diergeneeskunde en Dierenteelt* (NIBDD), vol. 36, 1924, p. 352.

74 Freijss, 'Schets van den Handel van Sumbawa', 1859, p. 268; and De Boer and Westermann, *Een Halve Eeuw Paketvaart*, 1941, p. 274.

75 't Hoen, *Veerassen en Veeteelt in Nederlandsch-Indië*, 1919, p. 18; Sibinga Mulder, 'De Economische Beteekenis van het Vee in Nederlandsch Oost-Indië en de Regeeringszorg Ervoor', 1927, p. 322; and *National Conservation plan for Indonesia. Volume IV: Nusa Tenggara*, Bogor: UNDP/FAO (Field Report no. 44, The National Parks Development Project, based on the work of John MacKinnon and others), 1982, p. 2.

76 Jan Merkens, 'De Grootveestapel van Nederlandsch-Indië' [The Livestock Population of the Netherlands-Indies], *Koloniale Studiën*, vol. 8 (Part I), 1924, pp. 189 and 195.

77 Pieter Hoekstra, 'Paardenteelt op het Eiland Soemba' [Horse-Breeding on the Island of Sumba], PhD thesis, Universiteit van Indonesië, Batavia, 1948, p. 23.

78 Kuperus, 'Beschouwingen over de Ontwikkeling en den Huidigen Vormenrijkdom van het Cultuurlandschap', 1938, p. 226; and Dove, 'Man, Land and

Game in Sumbawa', 1984, p. 116. *Alang-alang* is a species of high grass often known in English as elephant grass.

79 Jaarverslag van de BVD, 1926; Jaarverslag van de BVD, 1937; and Jaarverslag van de BVD, 1938.

80 Jaarverslag van de BVD, 1938; MMK 340, MvO, C. H. van Rietschoten, Timor en Onderhorigheden [Timor and Dependencies], 25.7.1913, p. 58.

81 Ajoebar, 'De Verbetering van den Paardenstapel op West-Soembawa', 1927, p. 367.

82 Glanders is an infectious disease of the mucous membrane of the nose, anthrax an acute infectious disease caused by the spore-forming bacterium *Bacillus anthracis* and trypanosomiases a kind of sleeping disease caused by the parasite *Trypanosoma evansi* and transmitted by biting flies (tsetse flies).

83 Jaarboek van het Departement van Landbouw, Nijverheid en Handel, 1911/ 1912, p. 262; and MMK 350, MvO, Groenevelt, afdeeling Soembawa [Department Sumbawa], 23.2.1927. After the formation of a veterinary service on Sumbawa in 1912, the vaccination of livestock and other preventive measures were undertaken. These measures proved to be highly effective and almost completely eliminated livestock diseases. KIT 1268, MvO, A. Couvreur, Timor en Onderhorigheden [Timor and Dependencies], 21.7.1924; MvO, Schultz, 1927, p. 13; and Ajoebar, 'De Verbetering van den Paardenstapel op West-Soembawa', 1927, p. 367.

84 MvO, Van Rietschoten, 1913, p. 58; Merkens, 'De Grootveestapel van Nederlandsch-Indië', 1923, p. 21; and De Boer and Westermann, *Een Halve Eeuw Paketvaart*, 1941, p. 275.

85 Population estimated on Jaarverslagen van de BVD, 1922–1938.

86 MvO, De Nijs Bik, 1934, p. 51.

87 De Boer and Westermann, *Een Halve Eeuw Paketvaart*, 1941, p. 276.

Chapter 5. Javanese Horses for the Court of Ayutthaya
Dhiravat na Pombejra

Currency

One *rijksdaalder* or rixdollar was equivalent to 60 *stuivers* in the East Indies after 1656. The value of the the Spanish *rial* of eight or *real* fluctuated, but was equivalent to 60 *stuivers* after 1622. In Siamese currency, one catty or *chang* was equivalent to 20 taels (*tamlung*), or 80 ticals (*baht*). Between 1665 and around 1690 one Siamese catty was equivalent to 144 Dutch guilders. Sources: George Vinal Smith, *The Dutch in Seventeenth-Century Thailand*, De Kalb: University of Northern Illinois Center for Southeast Asian Studies, 1977, pp. 134–135; *VOC-Glossarium*, Den Haag: Instituut voor Nederlandse Geschiedenis, 2000.

1 See for example Henry Ginsburg, *Thai Manuscript Painting*, London: The British Library, 1989, pp. 48, 65; Pamela York Taylor, *Beasts, Birds, and Blossoms in Thai Art*, Kuala Lumpur: Oxford University Press, 1994, p. 33. A *naga* is a mythical serpent which features in stories both of the Hindu god Vishnu and the Buddha. A famous representation in Thai art of horse and rider is of course the 'Frenchman' in the Lacquer Pavilion at Suan Pakkad Palace.

2 *Phra traibidok lae atthakatha thai* (Tripitaka), Mahamakut Buddhist University edition, Bangkok, 1982, 'Chakkawattisut' in 'Phra suttantabidok tikhanikai patikawak' Book 3 Part 1, pp. 99-100.

3 Nationaal Archief (formerly Algemeen Rijksarchief), Den Haag (henceforth NA), VOC 1517, King of Siam to Governor-General (hereafter G-G), BE 2235, fs.292 verso-293.

4 W. Ph. Coolhaas (ed.), *Generale Missiven van Gouverneurs-Generaal en Raden aan Heren XVII der Verenigde Oostindische Compagnie* [General Letters from the Governors-General and Council to the Gentlemen Seventeen of the United East India Company], vol. VIII, The Hague: Martinus Nijhoff, 1985, letter of 30 Nov. 1725, p. 26.

5 See Dhiravat na Pombejra, *Siamese Court Life in the Seventeenth Century as Depicted in European Sources*, Bangkok: Chulalongkorn University, 2001, pp. 157-159.

6 Armando Cortesão (ed.), *The Suma Oriental of Tomé Pires*, New Delhi & Madras: Asian Educational Services, 1990, vol. 1, pp. 201-203; M.C. Ricklefs, *A History of Modern Indonesia*, London and Basingstoke: Macmillan, 1981, p. 19.

7 D.G. Stibbe (ed.), *Encyclopaedie van Nederlandsch-Indië*, Tweede druk, Derde deel, The Hague, Leiden: Martinus Nijhoff/E. J. Brill, 1919, p. 228. The Java War was an unsuccessful rebellion by Prince Dipanagara and his followers against the Dutch.

8 De Graaff, cited by Denys Lombard, *Le sultanat d'Atjéh au temps d'Iskandar Muda 1607-1636*, Paris: Ecole Française d'Extrême Orient, 1967, p. 44.

9 See for instance Maurizio Bongianni (ed.), *Simon & Schuster's Guide to Horses and Ponies*, New York: Simon and Schuster, 1988, entries 126-130.

10 Even as late as the seventeenth century, the rulers of Southeast Asia were still searching for fine specimens of horse. In 1648 the *susuhunan* of Mataram sent three people to Persia to try to buy horses 'with wholly black, red, or white hair', and wanted the VOC to be of assistance to these horse-buyers. See Coolhaas (ed.), *Generale Missiven*, vol. II, 1964, letter of 18 Jan. 1649, p. 339.

11 John Crawfurd, *A Descriptive Dictionary of the Indian Islands & Adjacent Countries*, Kuala Lumpur: Oxford University Press, 1971, p. 382.

12 Joost Schouten, 'A Description of ... Siam' in François Caron and Joost Schouten, *A True Description of the Mighty Kingdoms of Japan and Siam* (facsimile of 1671 London edition), Bangkok: The Siam Society, 1986, p. 128.

13 Jeremias van Vliet, 'Description of the Kingdom of Siam', trans. L. F. van Ravenswaay, in Chris Baker et al (eds), *Van Vliet's Siam*, Chiang Mai: Silkworm, 2005; the Englishman William Keeling, in a journal entry for 22 November 1608, claims to have heard from a Siamese envoy in Banten that the Siamese clothed their horses and elephants in red cloths, presumably Indian cotton cloth. See Samuel Purchas, *Purchas His Pilgrims*, London: William Stansby, 1625, Book III, Chapter VI, p. 195.

14 Nicolas Gervaise, *The Natural and Political History of the Kingdom of Siam*, trans. John Villiers, Bangkok: White Lotus, 1989, p. 212; Simon de La Loubère, *The Kingdom of Siam*, Kuala Lumpur: Oxford University Press, 1969, p. 97. The 'Moors' traded with and in Siam from at least the sixteenth century, and by the early seventeenth century they, especially the Persians, had become very influential in the 'port department of the right' in Ayutthaya.

15 François-Timoléon, abbé de Choisy, *Journal of a Voyage to Siam 1685–1686*, trans. Michael Smithies, Kuala Lumpur: Oxford University Press, 1993, pp. 160–161.

16 Engelbert Kaempfer, *A Description of the Kingdom of Siam 1690*, trans. J.G. Scheuchzer, Bangkok: White Orchid Press, 1987, p. 45.

17 Lombard, *Le sultanat d'Atjéh*, p. 144; Lombard uses data from the *Hikayat Aceh*.

18 Jacques Dumarçay, *The Palaces of South-East Asia*, Singapore: Oxford University Press, 1991, p. 111.

19 Dhiravat, *Siamese Court Life*, pp. 96–97; Winai Pongsripian (ed.), *Kot monthianban chabab chaloem phra kiat* [The Palatine Law (Royal celebratory edition)], Bangkok: Thailand Research Fund, 2005, p. 97.

20 Van Vliet, 'Description', in Baker et al (eds), *Van Vliet's Siam*, p. 123.

21 La Loubère, *Kingdom of Siam*, p. 91. Apart from the huge numbers of horses involved in the wars of Europe from medieval times onwards, not to mention during the Classical era, large numbers of horses were also used in the wars in India during the seventeenth century. A VOC general letter of 1676 relates that the battles in the war involving Sivaji (Shivaji), the Maratha leader, against the Mughals and Bijapur involved up to 40,000 or even 70,000 horses, indeed quite different in scale from the 2,000 or so horses of King Narai's cavalry. See Coolhaas (ed.), *Generale Missiven*, , vol. IV, 1971, letter of 30 Sept. 1676, p. 119.

22 NA, VOC 1180, Van Muijden to G-G, 10 Nov. 1649, fs.604–611.

23 Ibn Muhammad Ibrahim Muhammad Rabi, *The Ship of Sulaiman*, trans. John O'Kane, London: Routledge and Kegan Paul, 1972, p. 85.

24 Krom Sinlapakon, *Prachum phongsawadan phak thi 82. Ruang phra ratcha phongsawadan krung siam chak ton chabap khong British Museum krung London* [British Museum recension of the Ayutthaya royal chronicles], Bangkok: Fine Arts Department, 1994, pp. 77–78, 121, 148.

25 G. E. Harvey, *History of Burma*, London: Frank Cass, 1967, p. 242.

26 Winai Pongsripian (ed.), *Khamhaikan khun luang wat pradu songtham* [Testimony of Khun Luang Wat Pradu Songtham], Bangkok: Secretariat of the Prime Minister's Office, 1991, p. 15.

27 NA, VOC 1691, Cleur's 'Relaas', 1703, fs.68-70.

28 Krom Sinlapakon, *Tamra ma khong kao kap tamra laksana ma* [old horse manuals], Bangkok: Fine Arts Department, 2001, pp. 1-6.

29 Krom Sinlapakon, *Krabuan phayuhayatra sathonlamak samai somdet phra naraimaharat chamlong chak ton chabab nangsu samud thai khong ho samud haeng chat* [Royal land procession in King Narai's reign from old Thai manuscript in National Library], Bangkok, 1987 and *Riu krabuan hae phayuhayatra thang chonlamak chak ton chabab nangsusamud thai khong ho samud haeng chat* [Water procession from King Narai's reign], Bangkok, 1987.

30 Remco Raben and Dhiravat na Pombejra (eds), *In the King's Trail*, Bangkok: Royal Netherlands Embassy, 1997, p. 17.

31 See also Krom Sinlapakon, *Tamra phichai songkhram* [Manual on warfare], Bangkok: Fine Arts Department, 1969, pp. 7, 25, 36-37.

32 NA, VOC 875, G-G to Okya Sombatthiban, 11 May 1651, f. 177.

33 NA, VOC 1377, Phrakhlang to G-G, 1682, fs.539-540.

34 Coolhaas (ed.), *Generale Missiven*, vol. IV, letter of 31 May 1684, p. 685; letter of 11 Dec. 1685, p. 837.

35 Coolhaas (ed.), *Generale Missiven*, vol. VII, 1979, letters of 14 Jan. and 28 Nov. 1715, pp. 149, 196-197.

36 Coolhaas (ed.), *Generale Missiven*, vol. VIII, letter of 5 Dec. 1726, p. 93.

37 La Loubère, *Kingdom of Siam*, pp. 39-40.

38 NA, VOC 1386, Phrakhlang to G-G, CS 1045/6, f.667 verso; VOC 1407, Okphra Kosathibodi to G-G & Council, 14 Dec. 1684, f.3223.

39 NA, VOC 8360, Kamer Zeeland, 'Overgegeven Gravaminas door den gesant van Siam aan haar Edes. tot Batavia' [List of grievances submitted by the Siamese envoy to the Governor-General and Council at Batavia], rec. 15 Feb. 1686; Coolhaas (ed.), *Generale Missiven*, vol. IV, letter of 31 May 1684, p. 687.

40 Coolhaas (ed.) *Generale Missiven*, vol. V, 1975, letter of 13 Dec 1686, p. 43.

41 Ibid, letter of 23 Dec. 1687, p. 126.

42 Coolhaas (ed.), *Generale Missiven*, vol. V, letter of 27 Dec. 1688, p. 217.

43 NA, VOC 1440, Keijts to G-G, 23 Dec. 1687, fs.2150 verso-2262 verso; Coolhaas (ed.), *Generale Missiven*, vol. V, letter of 13 March 1688, p. 171; Ricklefs, *History*, pp. 80-81.

44 On the Dutch in Siam after 1688, see for example George Vinal Smith, *The Dutch in Seventeenth-Century Thailand*, De Kalb: University of Northern Illi-

nois Center for Southeast Asian Studies, 1977, p. 45; Han ten Brummelhuis, *Merchant, Courtier and Diplomat. A History of the Contacts between the Netherlands and Thailand*, Lochem-Gent: De Tijdstroom, 1987, pp. 40 ff.

45 NA, VOC 1503, Van Son to G-G, 23 Dec. 1692, fs.531 & verso.

46 NA, VOC 1485, Van den Hoorn to G-G & Council, 5 Feb. 1690, f.120; VOC 1498, King Phetracha to G-G, 1690-1691, f.248 verso;VOC 1517, Phrakhlang to G-G & Council, received 17 March 1692.

47 NA, VOC 1596, Phrakhlang to G-G & Council, received 18 Feb. 1697, fs.2-4. A cuirassier was a mounted soldier wearing metal or leather armour (the 'cuirass'), covering the chest and sometimes the back. In Europe the term came into use during the mid-sixteenth century.

48 NA, VOC 1569, Phrakhlang to G-G & Council, BE 2238, received 18 Jan. 1695, fs.56-65; VOC 1536, King of Siam to G-G & Council, received 7 Dec. 1693, fs.88 & verso; Krom Sinlapakon, *Tamra ma*, pp. 1-2.

49 NA, VOC 1517, Phrakhlang to G-G & Council, BE 2236, f.427; VOC 1569, Phrakhlang to G-G & Council, received 18 Jan. 1695, f.99.

50 NA, VOC 1517, Phrakhlang to G-G & Council, received 17 March 1692; VOC 1623, Tant to G-G & Council, 6 Jan. 1699, fs.10-11. In the traditional Siamese administrative hierarchy, the ranks conferred on officials were (in descending order of importance) *chaophraya*, *okya* or *phraya*, *okphra* or *phra*, *okluang* or *luang*, *okkhun* or *khun*, *okmun* or *mun*, and *okphan* or *phan*.

51 NA, VOC 1663, Tant & Council to G-G & Council, 31 Jan. 1702, fs.26-27.

52 The titles of the people sent to Java to buy horses rarely corresponded to titles in the lists included in the Laws of Hierarchy, probably because those lists were by no means comprehensive. An exception was 'Coen Craij sinthop', which corresponds exactly with the rank and title of *khun krai sinthop*, an official in the royal horses department or *krom ma ton* with 400 *sakdi na* marks. The head of this department was the *okya suriya phaha*, with 3,000 *sakdi na* marks. See Krom Sinlapakon, *Ruang kotmai tra sam duang* [The Law of the Three Seals], Bangkok: Fine Arts Department, 1978, pp. 131-133.

53 Coolhaas (ed.), *Generale Missiven*, vol. V, letter of 19 Jan. 1697, pp. 787-788.

54 NA, VOC 1506, Van Son & Council to G-G & Council, 8 Dec. 1697, f.54.

55 NA, VOC 1569, Phrakhlang to G-G & Council, received 18 Jan. 1695, f.107; Phrakhlang to G-G & Council, received 18 Feb. 1697, fs.2-4; Coolhaas (ed.), *Generale Missiven*, vol. V, letter of 19 Jan. 1697, p. 788; letter of 30 Nov. 1697, p. 842; *Generale Missiven*, vol. VI, 1976, letter of 2 Feb. 1698, p. 5.

56 NA, VOC 1676, Tant & Council to G-G & Council, 1 Feb. 1703, fs.93-94; VOC 1676, Cleur & Council to G-G & Council, 8 Jan. 1704, f.20; VOC 939, G-G & Council to Cleur & Council, 16 Aug. 1703.

57 NA, VOC 939, G-G & Council to Cleur & Council, 16 Aug. 1703, f.727; see also Bhawan Ruangsilp, *Dutch East India Company Merchants at the Court of Ayutthaya: Dutch Perceptions of the Thai Kingdom, c. 1604-1765,* Leiden/Boston: Brill, 2007, pp. 171-172.

58 Société des Missions Etrangères, rue du Bac, Paris (Archives), vol. 882, 'Relation de la Mission de Siam' by Gabriel Braud, fs.18-22.

59 NA, VOC 1569, Phrakhlang to G-G & Council, BE 2238, received 18 Jan. 1695, fs.61-64.

60 See for example David K. Wyatt, *Thailand: A Short History,* London and Bangkok: Yale University Press, 1984, pp. 126-127.

61 Ten Brummelhuis, *Merchant,* pp. 44-45; Coolhaas (ed.), *Generale Missiven,* vol. VI, letter of 25 Nov. 1708, p. 544; letter of 30 Nov. 1709, p. 610.

62 Coolhaas (ed.) *Generale Missiven,* vol. VI, letter of 19 Feb. 1710, p. 659; letter of 30 Nov. 1710, p. 681.

63 J. van Goor (ed.), *Generale Missiven,* vol. IX, The Hague: Martinus Nijhoff, 1988, letter of 17 Oct. 1730, p. 114; letter of 30 Nov. 1730, p. 174.

64 Coolhaas (ed.), *Generale Missiven,* vol. II, letter of 8 Jan. 1641, p. 132.

65 Coolhaas (ed.), *Generale Missiven,* vol. III, 1968, letter of 19 Dec. 1671, p. 764.

66 Van Goor (ed.), *Generale Missiven,* vol. IX, letter of 14 Feb. 1732, p. 299; letter of 28 Dec. 1731, p. 266; letter of 12 Oct. 1731, pp. 235-236; letter of 5 Dec. 1726, p. 79.

67 NA, VOC 964, G-G & Council to D. Blom & Council, 24 July 1716, f.746.

68 NA, VOC 970, G-G & Council to W. Blom & Council, 18 Aug. 1719, fs.813-814.

69 Data on loans to Siamese horse-buyers from the *Generale Missiven* series, edited by Coolhaas and Van Goor, vols V-IX.

70 Van Goor (ed.), *Generale Missiven,* vol. IX, letter of 8 Dec. 1732, p. 635; letter of 12 Oct. 1735, p. 654; letter of 17 Oct. 1730, p. 114; NA, VOC 2346, Name roll of the King of Siam's servants travelling per the *Jacoba,* 31 Jan. 1735, fs.41-42; J.E. Schooneveld-Oosterling (ed.), *Generale Missiven,* vol. XI, The Hague: Martinus Nijhoff, 1997, letter of 31 December 1748, p. 656.

Chapter 6. Colonising New Lands: Horses in the Philippines
Greg Bankoff

Earlier versions of this chapter can be found in Greg Bankoff, '*Bestia Incognita*: The Horse and Its History in the Philippines 1880-1930', *Anthrozoös,* vol. 17, no. 1, 2004, pp. 3-25; and Greg Bankoff, 'Horsing Around: The Life and Times of the Horse in the Philippines at the Turn of the 20th Century'. In Peter Boomgaard and

David Henley (eds), *Smallholders and Stockbreeders: Histories of Foodcrop and Livestock Farming in Southeast Asia*, Leiden: KITLV Press, 2004, pp. 233-255.

1   This ancestry is still hinted at etymologically by the use of the Malay word *kuda* for horse in Mindanao and Sulu languages. William Scott, *Barangay: Sixteenth Century Philippine Culture and Society*, Quezon City: Ateneo de Manila, 1994, p. 278, fn.15.

2   Felipe II, Instructions from Felipe II to Governor Gómez Pérez Dasmariñas, 9 August, 1589. In Emma Blair and Alexander Robertson, *The Philippine Islands, 1493-1898*, Mandaluyong: Cachos Hermanos, vol. 7, 1973, p. 156; Pedro Chirino, 'Relación de las Islas Filipinas, 1604'. In Blair and Robertson, *The Philippine Islands, 1493-1898*, vol. 12, 1973, p. 191; and 'Animal Industry in the Philippines; Horses', *Philippine Agricultural Review*, vol. 4, no. 9, 1911, pp. 476-478. The island of Timor is also mentioned as a source of supply. Frederick Wernstedt and J. E. Spencer, *The Philippine Island World: A Physical, Cultural and Regional Geography*, Berkeley: University of California Press, 1978, p. 211.

3   William Dampier, 'A New Voyage Round the World, 1703'. In Blair and Robertson, *The Philippine Islands, 1493-1898*, vol. 38, 1973, p. 282 and vol. 39, 1973, p. 87; and Nicholas Nicols, 'Commerce of the Philippine Islands, 1759'. In Blair and Robertson, *The Philippine Islands, 1493-1898*, vol. 47, 1973, p. 302.

4   Greg Bankoff, Chapter 7, this volume.

5   'Remarks on the Philippine Islands, and on their Capital Manila, 1819 to 1822, By an Englishman, 1828'. In Blair and Robertson, *The Philippine Islands, 1493-1898*, vol. 51, 1973, p. 128.

6   'Memoria Preliminar para la Ejecución del Proyecto de Renovación y Mejora de la Cria Caballar de Este Archipiélago' [Preliminary Report on the Management of the Project to Reinvigorate and Improve the Breed of Horse of this Archipelago], Philippine National Archives (hereafter PNA), Raza de Caballeria, Formento y Mejora, 1883.

7   This task is still often fulfilled by horse-carts in provincial cities and even in Metro Manila.

8   Report of the Philippine Commission to the President, January 31, 1900, Washington: Government Print Office, vol. 4, 1900, p. 80; and José Montero y Vidal, *El Archipiélago Filipino y las Islas Marianas, Carolinas y Palaos: Su Historia, Geografía y Estadística* [The Philippine Archipelago and the Marianas, Carolinas and Palau Islands: Their History, Geography and Statistical Profile] Madrid: Imprenta y Fundición de Manuel Tello, 1886, p. 285. On roadways in the Philippines, see Maria Isable Piquerras Villaldea, *Las Comunicaciones en Filipinas durante el Siglo XIX: Caminos, Carreteras y Puentes* [Communi-

cations in the Philippines During the 19th Century: Roads, Highways and Bridges], Madrid: Archiviana, 2002.

9   An *arroba* is equivalent to approximately 11.5 kilograms.

10 'Expediente sobre Dejar Sin Efecto la Circular de Este Centro, de 27 de Julio de 1886' [Proceedings on Leaving in Effect the Circular of this Centre, of 27 July 1886], PNA, Carruajes, Carros y Caballos, Bundle 7, 1887.

11 'Supresión del Impuesto sobre los Caballos, Abra 1889' [Abolition of the Tax on Horses, Abra 1889], PNA, Carruajes, Carros y Caballos, Bundle 7, 1889; Blas Jerez to Director-General de Administración Civil, PNA, Carruajes, Carros y Caballos, Bundle 7, 1889; and Maximo Loilla to Director-General de Administración Civil, PNA, Carruajes, Carros y Caballos, Bundle 7, 1889.

12 'Memoria de Ramon', PNA, Servicios de Agricultoras, Cebu 1890–97, 1893; and 'Memoria sobre un Tranvia, Cebu', PNA, Servicios de Agricultoras, Cebu 1890–97, 1893.

13 John Foreman, *The Philippine Islands*, Mandaluyong: Cacho Hermanos, 1985 (originally published 1899), p. 355; Marcelino Foronda, *Insigne y Siempre: Leal Essays on Spanish Manila*, Manila: De La Salle University, History Department and the Research Center, 1986, p. 135; and Montero y Vidal, *El Archipiélago Filipino*, 1886, p. 157.

14 Benito Legarda, *After the Galleons: Foreign Trade, Economic Change and Entrepreneurship in the Nineteenth-Century Philippines*, Quezon City: Ateneo de Manila, 1999, pp. 286, 329.

15 It is sometimes quite impossible to differentiate between saddle and pack use of horses as they might transport goods to market but be ridden back after their loads had been sold. Such definitional niceties were the occasion of much dispute as to whether these animals should be excluded from the taxes levied on the non-agricultural use of horses. 'Expediente sobre Reclamación, Iloilo 1884' [Proceedings about the Complaint, Iloilo 1884], PNA, Carruajes, Carros y Caballos, Bundle 7, 1884.

16 'Carruajes "La Union"' [Carriages "La Union"], PNA, Carruajes, Carros y Caballos, Bundle 7, 1889.

17 'Formación Previa en Averiguación' [Review of the Case], PNA, Veterinarios, 1874.

18 *Report of the Philippine Commission, 1903*, Washington: Government Printing Office, vol. 6, 1904, p. 423; and *Report of the Philippine Commission, 1905*, Washington: Government Printing Office, vol. 2, 1906, p. 459.

19 'Estado del Numero de Carruages, Carros y Caballos, Varias Provincias', 1889; and 'Relación de los Caballos' [Pertaining to Horses], San Pascual 10 February 1889', PNA, Carruajes, Carros y Caballos, Bundle 7, 1889.

20 Remarks on the Philippine Islands, 1973, vol. 51, p. 128; José Maria Clotet, Fr José Maria Clotet to Rev. Fr Rector of Ateneo Municipal, 11 May, 1889. In Blair

and Robertson, *The Philippine Islands, 1493-1898*, vol. 43, 1973, p. 160; and Charles Wilkes, *Narrative of the United States Exploring Expedition during the Years 1838, 1839, 1840, 1841, 1842*. In Blair and Robertson, *The Philippine Islands, 1493-1898*, vol. 43, 1973, p. 160. On the Wilkes Expedition, see William Stanton, *The Great United States Exploring Expedition of 1838-1842*, Berkeley, Los Angeles and London: University of California Press, 1975.

21 José Montero y Vidal, *Historia General de Filipinas, desde el Descubrimiento de Dichas Islas Hasta Nuestros Días* [A General History of the Philippines, Since the Discovery of the Said Islands until Nowadays], Madrid: Establecimiento Tipográfico de la Viuda é Hijos de Tello, 1895, vol. 3, p. 496.

22 'Espediente Original, Pampanga 1859' [The First or Original Proceedings], PNA, Carreras de Caballos, Bundle 1, 1851; and 'Pliego de Condiciones de Carrera de Caballo de la Provincia de la Pampanga' [Details of the Main Horse Track of the Province of Pampanaga], PNA, Carreras de Caballos, Bundle 1, 1879.

23 *The Manila Jockey Club: 130 Years of Horse Racing in Southeast Asia*, Manila: Manila Jockey Club, 1997, pp. 54-60; and Filomeno Aguilar, *Clash of Spirits: The History of Power and Sugar Planter Hegemony on a Visayan Island*, Honolulu: University of Hawai'i Press, 1998, p. 192. Two new clubs were founded in the 1930s, one in Cebu that proved a rather short-lived affair and the Philippine Racing Club in Santa Ana, Manila that continues to hold races today. On gambling, see Aguilar, *Clash of Spirits*, 1998, pp 32-62; and Greg Bankoff, 'Redefining Criminality: Gambling and Financial Expediency in the Colonial Philippines 1764-1898', *Journal of Southeast Asian Studies*, vol. 22, 1991, pp. 267-281.

24 *The Manila Jockey Club: 130 Years of Horse Racing*, 1997, p. 55.

25 'Live Stock and Poultry in the Philippines; Horses', *The Philippine Agricultural Review*, vol. 4, 1911, p. 481.

26 'Una Epizootia en Filipinas' [An Epizootic in the Philippines], PNA, Veterinarios, 1888; Marshall McLennan, *The Central Luzon Plain: Land and Society on the Inland Frontier*, Quezon City: Alemar-Phoenix Publishing House, 1980, p. 169; and Reynaldo Ileto, 'Hunger in Southern Tagalog, 1897-1898'. In Reynaldo Ileto, *Filipinos and their Revolution: Event, Discourse, and Historiography*, Quezon City: Ateneo de Manila Press, 1998, pp. 113-115.

27 'Una Epizootia en Filipinas', 1888.

28 *Report of the Philippine Commission, 1903*, vol. 6, 1904, pp. 421, 469; M. Mitzmain, 'Collected Notes on the Insect Transmission of Surra in Carabaos', *Philippine Agricultural Review*, vol. 5, no. 12, 1912, p. 679; and David Kretzer, 'How to Build Up and Improve a Herd or Flock (With Description of their Most Common Diseases in the Philippines)', *Philippine Agricultural Review*, vol. 21, 1928, pp. 261, 262.

29 *Report of the Philippine Commission, 1903*, vol. 6, 1904, p. 421.

30 'Surra', *Philippine Agricultural Review*, vol. 1, 1908, p. 119.

31 *Census of the Philippine Islands, 1903*, Washington: Bureau of the Census, vol. 4, 1905, p. 248; and *Census of the Philippine Islands, 1918*, vol. 3, 1921, p. 743.

32 The animal epidemics that decimated herds throughout the archipelago caused untold hardship to Filipinos in terms of both agricultural productivity and overland mobility. Accounts of the war barely mention the former and completely ignore the latter, when the importance of the small native horse that provided the chief means of land transportation in the islands was enhanced by an American naval blockade that restricted sea-borne communications. For an example of this omission, see Brian McAllister Linn, *The Philippine War 1899–1902*, Lawrence, Kansas: University of Kansas Press, 2000.

33 For a study of the effects of natural hazards on the Philippines, see Greg Bankoff, *Cultures of Disaster: Society and Natural Hazard in the Philippines*, Richmond: RoutledgeCurzon, 2003.

34 'Floods in the Philippines 1691–1911', *Archives of the Manila Observatory*, Box 10, 37.

35 Montero y Vidal, *El Archipiélago Filipino*, 1886, pp. 155–156, 329, 342, 361, 437; and *Census of the Philippine Islands, 1918*, 1921, vol. 3, p. 734.

36 Estadística Demografico-Sanitaria del Radio Municipal de L. M. I. Y. S. L.C. de Manila de Enero a Diciembre de 1886, Defunciones [Demographic and Health Statistics of the Municipal Authority of Manila from January to December 1886, Deaths], *El Comercio*, 3 March, 1888; 'Estado del Numero de Carruages, Carros y Caballos, Varias Provincias', 1889.

37 Commander of the GCV to the Director-General of Civil Administration, 4 September, PNA, Veterinarios, 1888.

38 José Sora to Sor. Corregidor de Manila, PNA, Animales Sueltos, Corregimiento de Manila, 1878; El Comandante Jefe to Corregidor de Manila, PNA, Animales Sueltos, Corregimiento de Manila, 1884; 'Yncidente sobre la Detención de un Caballo' [Incident to do with Detaining a Horse], PNA, Animales Sueltos, Corregimiento de Manila, 1884; and Manuel de Vos to Gobernador Civil de Provincia de Manila, PNA, Animales Sueltos, Corregimiento de Manila, 1896.

39 'La Aprehensión de 23 Individuos de la Provincia de la Laguna y Cavite, Manila, 15 June 1891' [The Detention of 23 Suspects from the Provinces of Laguna and Cavite, Manila, 15 June 1891], PNA, Cuadrilleros, Bundle 1, 1891.

40 Francisco de Viana, Memorial of 1765. In Blair and Robertson, *The Philippine Islands, 1493–1898*, vol. 48, 1973, p. 305. Horse-flesh was also eaten to a limited extent in other parts of Southeast Asia especially in the Batak, Gayo and Minangkabau regions of Sumatra. William Gervase Clarence-Smith,

'Horse Breeding in Mainland Southeast Asia and Its Borderlands'. In Peter Boomgaard and David Henley (eds), *Smallholders and Stockbreeders: Histories of Foodcrop and Livestock Farming in Southeast Asia*, Leiden: KITLV Press, 2004, pp. 189–210; and William Gervase Clarence-Smith, Chapter 2, this volume.

41 'Sobre los Gravisimos Perjuicios, 30 May 1807' [Concerning the Severest of Wrongs], PNA, Spanish Manila, Reel 5, 1807; and Alfred Marche, *Luzon and Palawan*, Manila: Filipiniana Book Guild, 1968 (originally published 1905), p. 75.

42 W. Best, 'Suggestions for the Care of Horses in the Philippine Islands', *Philippine Agricultural Review*, vol. 7, 1914, pp. 330–331.

43 'Una Epizootia en Filipinas', 1888; and Kretzer, 'How to Build Up and Improve a Herd or Flock', 1928, pp. 262–263, 264.

44 'Tarifa de los Honorarios' [Price of Services], PNA, Veterinarios, 1864; and Kretzer, 'How to Build Up and Improve a Herd or Flock', 1928, pp. 266–280.

45 'Tarifa de los Honorarios', 1864; and Felipe Jovantes to Guardia Civil Veterana, PNA, Veterinarios, 1874.

46 *Report of the Philippine Commission, 1904*, Washington: Government Printing Office, 1905, vol. 2, p. 205–206; and *Report of the Philippine Commission, 1913*, Washington: Government Printing Office, 1914, pp. 299–301.

47 'Quarantine Stations', *Philippine Agricultural Review*, vol. 3, 1910, p. 27; and D. Palmer and Victor Buencamino, 'The College of Veterinary Science, University of the Philippines', *Philippine Agricultural Review*, vol. 6, 1913, pp. 368–370. On the history of veterinary medicine in the Philippines, see Mauro Manuel et al (eds), *A Century of Veterinary Medicine in the Philippines*, Quezon City: University of the Philippines Press, 2002.

48 The category of 'other' is undefined but presumably includes death from accident and misadventure.

49 'Testimonio 22 June 1849' [Testimony of 22 June 1849], PNA, Matanzas y Limpieza de Reses, 1849; 'Una Epizootia en Filipinas', 1888; 'Pliego de la Matanza y Limpieza de Reses, Fecha 4 de Mayo de 1880, Manila 1 March 1889' [Folder on the Slaughtering and Cleaning of Cattle, 1889], PNA, Matanzas y Limpieza de Reses, 1889; and 'Live Stock and Poultry in the Philippines; Horses', 1911, p. 482.

50 *Report of the Philippine Commission to the President, January 31, 1900*, vol. 4, 1900, p. 13.

51 Montero y Vidal, *El Archipiélago Filipino*, 1886, pp. 155–156, 329, 342, 361, 437.

52 *Census of the Philippine Islands, 1918*, vol. 3, 1921, pp. 734–735. The 1918 Census of the Philippine Islands provides provincial population figures for

both 1903 and 1918. Moreover, the Southern Tagalog region comprising Morong (Rizal), Laguna, Cavite, Tayabas as well as Batangas accounted for nearly 26 per cent of all domesticated horses in 1886. Montero y Vidal, *El Archipiélago Filipino*, 1886, pp. 155-156, 329, 342, 361, 437.

53 De Jong Boers, Chapter 4, this volume.

54 Montero y Vidal, *El Archipiélago Filipino*, 1886, pp. 155-156, 329, 342, 361, 437. A comparison of domestic animal populations between 1886 and 1911 shows an enormous increase in the numbers of pigs (2,066,000), goats (407,000) and sheep (88,000), and a corresponding decrease in carabaos (713,000) and beef cattle (242,000) numbers, the decrease the result of the rinderpest epidemic in the late nineteenth century. Horse numbers remain largely unchanged (215,000). Manuel Avellano, *Geografía de las Islas Filipinas*, Manila: Tipografía del Colegio de Sto. Tomás, 1912, p. 46.

55 Bernice De Jong Boers, 'Paardenfokkerij op Sumbawa (1500-1930)' [Horse Breeding in Sumbawa], *Spiegel Historiael*, vol. 10/11, no. 32 October/November 1997, pp. 438-443; and de Jong Boers, Chapter 4, this volume.

56 The recommended dietary regime suggested that a quarter of the animal's feed be given in the morning, the same at midday and the remainder in the evening. A two-hour period was then allowed to digest this meal. It was mistakenly assumed that horses required a longer period for digestion than carabaos or bullocks and so could not be put to work straight after eating. In fact, the opposite is true: unlike other common domesticated ungulates such as cattle, sheep and goats, the horse breaks down the cellulose in fibrous plants through a cecal digestive system rather than in a rumen and can literally eat and run at the same time. Older animals were thought to require more food as their teeth were deemed to be inefficient grinders and so to waste much of the feed's nutritional value. Based on this dietary regime, a healthy horse was then expected to work a nine-hour day, two stints of four and a half hours broken by the midday rest period. Best, 'Suggestions for the Care of Horses in the Philippine Islands', 1914, pp. 328-329.

57 De Jong Boers, 'Paardenfokkerij op Sumbawa (1500-1930)', 1997, p. 441; and Best, 'Suggestions for the Care of Horses in the Philippine Islands', 1914, pp. 328-329.

58 E. Koert, 'Stock Breeding in the Catanduanes Islands', *Philippine Agricultural Review*, vol. 5, 1912, p. 306.

59 Antonio de Morga, 'Sucesos de las Islas Filipinas, 1609'. In Blair and Robertson, *The Philippine Islands, 1493-1898*, vol. 16, 1973, pp. 90-91; and *Report of the Philippine Commission, 1905*, vol. 2, 1906, p. 459.

60 De Jong Boers describes the practice of setting fire to the *alang-alang* or imperata grass on Sumbawa to create additional pasturage. De Jong Boers, 'Paardenfokkerij op Sumbawa (1500-1930)', 1997, p. 440. As yet no evidence

of a similar technique has been noted in the Philippines though this more likely reflects lack of research.

61 'Proyecto de Renovación y Mejora de la Cria Caballar' [Project to Reinvigor-ate and Improve the Breed of Horse of this Archipelago], PNA, Raza de Caballeria de Filipinas, Formento y Mejora, 1881; and 'Memoria Preliminar para la Ejecución del Proyecto de Renovación y Mejora de la Cria Caballar de Este Archipiélago' [Preliminary Report on the Management of the Project to Reinvigorate and Improve the Breed of Horse of this Archipelago], 1883.

62 'Memoria Preliminar para la Ejecución del Proyecto de Renovación y Mejora de la Cria Caballar de Este Archipiélago', 1883.

63 Eleanor Melville, *A Plague of Sheep: Environmental Consequences of the Con-quest of Mexico*, Cambridge: Cambridge University Press, 1994; and Greg Bankoff and Sandra Swart, Chapter 1, this volume.

64 'Memoria Preliminar para la Ejecución del Proyecto de Renovación y Mejora de la Cria Caballar de Este Archipiélago', 1883; De Jong Boers, 'Paardenfokkerij op Sumbawa (1500-1930)', 1997, p. 441; and Clarence-Smith, 'Horse Breed-ing in Mainland Southeast Asia', 2004.

Chapter 7. Adapting to a New Environment: The Philippine Horse
Greg Bankoff

An earlier version of this chapter appeared as Greg Bankoff, 'A Question of Breeding: Zootechny and Colonial Attitudes Towards the Tropical Environment in Late Nineteenth Century Philippines', *Journal of Asian Studies*, vol. 60, no. 2, 2001, pp. 413-437 and Greg Bankoff, '*Bestia Incognita*: The Horse and Its History in the Philippines 1880-1930', *Anthrozoös*, vol. 17, no. 1, 2004, pp. 3-25.

1 Greg Bankoff, Chapter 6, this volume.

2 The animals appear slightly coarser in build, with a longer and more massive head with some individuals retaining a slight preorbital depression, large first premolars and only 17 pairs of ribs that were characteristic of the Sulu Horse. [No author], 'Animal Industry in the Philippines; Horses', *Philippine Agricultural Review*, vol. 4, no. 9, 1911, p. 479. It was also generally con-sidered to have 'less style, action, and vigor than the northern pony'. C. Edwards, 'The Live-stock Industry of the Philippines', *Philippine Agricultural Review*, vol. 9, no. 2, 1916, p. 140.

3 [No author], 'Animal Industry', 1911, p. 480.

4 Antonio de Morga, 'Sucesos de las Islas Filipinas [1609]'. In Emma Blair and Alexander Robertson (eds), *The Philippine Islands, 1493-1898*, Mandaluyong: Cachos Hermanos, 1973, vol. 16, pp. 90-91.

5 See Sandra Swart, Chapter 8, this volume.

6   'Early Franciscan Missions [1649]'. In Blair and Robertson (eds), *The Philippine Islands, 1493-1898*, 1973, vol. 35, p. 299; 'Remarks on the Philippine Islands, and on their Capital Manila, 1819 to 1822 By an Englishman [1828]'. In Blair and Robertson (eds), *The Philippine Islands, 1493-1898*, 1973, vol. 51, p. 128; and Estevan Vivít, *Reseña Estadística de las Islas Filipinas en 1845* [Current Statistics of the Philippine Islands in 1845], Barcelona: Imprenta de A. Brosi, no date, p. 30.

7   William Gervase Clarence-Smith, 'Horse Breeding in Mainland Southeast Asia and Its Borderlands'. In Peter Boomgaard and David Henley (eds), *Smallholders and Stock-breeders: Histories of Foodcrop and Livestock Farming in Southeast Asia*, Leiden: KITLV Press, 2004, pp. 189-210; Peter Boomgaard, Chapters 3, this volume; Bernice de Jong Boers, Chapter 4, this volume.

8   Bernice de Jong Boers, 'Paardenfokkerij op Sumbawa (1500-1930)' [Horse Breeding in Sumbawa], *Spiegel Historiael*, vol. 10/11, no. 32 October/November, 1997, p. 441; and De Jong Boers, Chapter 4, this volume.

9   'Expediente sobre la Ejecución del Proyecto de Renovación y Mejora de la Cria Caballar de este Archipiélago' [Proceedings on the Management of the Project to Reinvigorate and Improve the Breed of Horse of this Archipelago], Philippine National Archives (hereafter PNA), Raza de Caballeria, Formento y Mejora, 1881.

10 'Memoria Preliminar para la Ejecución del Proyecto de Renovación y Mejora de la Cria Caballar de Este Archipiélago'[Preliminary Report on the Management of the Project to Reinvigorate and Improve the Breed of Horse of this Archipelago], PNA, Raza de Caballeria, Formento y Mejora, 1883.

11 José Aguilar, *Colonización de Filipinas* [The Colonisation of the Philippines], Madrid: Establecimiento Tipográfico de Alfredo Alonso, 1893, p. 71; and 'Expediente sobre la Ejecución del Proyecto de Renovación y Mejora de la Cria Caballar', 1881.

12 'Memoria Preliminar para la Ejecución del Proyecto de Renovación y Mejora de la Cria Caballar', 1883.

13 Ibid. The standard contemporary measurement for horses is a 'hand' equivalent to four inches or 10.16 cm and measured at the shoulder. Modern horses are regarded as small between 10-12 hands (1.02-1.22 metres) in height, medium around 15-16 hands (1.52-1.63 metres), and large over 16 hands (1.63 metres). Spanish measurements are calculated from data given on horse measurements in both *cuartas* and metric calibrations. 'Inventario de los Cinco Caballos Arabes del Estado que Existen en este Gobierno que por Orden del Exmo Sor. Gobor. Gral. del Archipielago se Remiten a Disposicion del Ecmo. Sor. Director Gral. de Admon. Civil por Conducto del Vapor 'Elcano', 12 February' [Record of the Five Arab Stallions Currently in the Possession of the State that by Order of the Governor-general of this Archipelago are

placed at the Disposition of the Director-general of Civil Administration by Means of the Steamship 'Elcano'], PNA, Raza de Caballeria de Filipinas, Formento y Mejora, 1890.

14 Source for Figure 7.1 Nemesio Catalan, 'The Animal Problem of the Future Philippine Army', *Philippine Journal of Animal Industry*, vol. 2, no. 1, 1935, p. 70.

15 Measurements from 722 (wrongly stated as 772) animals were compiled and the proportion extrapolated and then applied to the estimated horse population of 350,000.

16 The minimum height requirement for riding horses in the U.S. Army was 58 inches though it was considered that beasts of 50 inches and above would have to serve as infantry machine gun transports in the Philippines through force of circumstances. Catalan, 'The Animal Problem of the Future Philippine Army', 1935, pp. 71-72.

17 Aguilar, *Colonización de Filipinas*, 1893, p. 260.

18 *Report of the Philippine Commission to the President, January 31, 1900*, Washington: Government Print Office, 1900, vol. 3, p. 308.

19 Several fish species in the Zuiderzee, the inland sea of the Netherlands, have also diminished in size since the completion of the *Afsluitdijk* in 1932 that separated it from the North Sea and effectively began the process that is turning its previously saline waters into a freshwater lake, the Ijsselmeer.

20 Bankoff, Chapter 6, this volume.

21 The British in Lesotho pursued a similar policy; see Sandra Swart, Chapter 9, this volume.

22 The five members of the subcommittee included Lieutenant Colonel of Artillery Arturo de Molins, Captain Commander of Cavalry José Sacarte, Veterinary Professor of Cavalry Ginés Geis, Brigadier of Engineers Felipe Lacorte and Brigadier of Cavalry José Sanchez Mira. There is some ambiguity over the year in which this committee was actually convened: one document dated September 1883, though admittedly a copy, refers to it being established on 23 April of that year (Raza de Caballeria de Filipinas 1883), but a second document that appears to be dated June 1881 clearly makes reference to the same committee's recommendations (Raza de Caballeria de Filipinas 1881). Yet a third document indicates that the second document was, in fact, drawn up in 1884. Respectively 'Memoria Preliminar para la Ejecución del Proyecto de Renovación y Mejora de la Cria Caballar', 1883; 'Expediente sobre la Ejecución del Proyecto de Renovación y Mejora de la Cria Caballar', 1881; and 'Expediente sobre Cumplimiento de la Real Orden No 275 De 30 Marzo de 1885 para el Fomento y Meyora de la Raza Caballar en las Islas, 1 May' [Proceedings on the Implementation of the Royal Order No.275 of 30 March 1885 to Promote and Improve the Breed

of Horse in the Islands], PNA, Raza de Caballeria de Filipinas, Formento y Mejora, 1888.

23 'Memoria Preliminar para la Ejecución del Proyecto de Renovación y Mejora de la Cria Caballar', 1883.

24 Manuel Avellano, *Geografía de las Islas Filipinas*, Manila: Tipografía del Colegio de Sto. Tomás, 1912, p. 52. This attempt was no more successful than previous Spanish endeavours with many of the horses falling victim to surra, glanders and other diseases and the animals adversely affected by ticks, flies and poor pasturage. Carlos Diaz, 'Breeding Performance of Arabian Stallions of the Bureau of Animal Industry', *Philippine Journal of Animal Industry*, vol. 3, no. 4, 1936, pp. 263-264.

25 'Expediente sobre la Ejecución del Proyecto de Renovación y Mejora de la Cria Caballar', 1881.

26 Thomas Glick, 'Spain'. In Thomas Glick (ed.), *The Comparative Reception of Darwin*, Austin and London: University of Texas Press, 1974, pp. 307-308.

27 'Expediente sobre la Ejecución del Proyecto de Renovación y Mejora de la Cria Caballar', 1881.

28 Montero y Vidal mentions the establishment of a *remonta* in January 1860 specifically to improve the breed of horses for the regiment of lancers but provides no other information or reference to this or any subsequent venture. José Montero y Vidal, *Historia General de Filipinas, desde el Descubrimiento de Dichas Islas Hasta Nuestros Días*, Madrid : Establecimiento Tipográfico de la Viuda é Hijos de Tello, 1895, vol. 3, p. 269.

29 'Expediente sobre la Ejecución del Proyecto de Renovación y Mejora de la Cria Caballar', 1881; and 'Memoria Preliminar para la Ejecución del Proyecto de Renovación y Mejora de la Cria Caballar', 1883.

30 'Memoria Preliminar para la Ejecución del Proyecto de Renovación y Mejora de la Cria Caballar', 1883.

31 'Memoria Preliminar para la Ejecución del Proyecto de Renovación y Mejora de la Cria Caballar', 1883. Costs included a purchase price (150 pesos), transportation (90 pesos), maintenance during acclimatisation (40 pesos) and contingencies (20 pesos).

32 'Expediente sobre Cumplimiento de la Real Orden No 275 De 30 Marzo de 1885 para el Fomento y Meyora de la Raza Caballar en las Islas, 1 May' 1888.

33 Investor confidence was dented by the mercantile crisis and repeated agricultural calamities, especially the rinderpest epidemic that affected the colony during the 1880s. Bankoff, Chapter 6, this volume. The decision to buy more stallions than mares in India reflects the Commission's views about the relative superiority of the female over the male horse in the Philippines as well as the practicalities of destroying an estimated existing 90,000 mares and the wealth they represented. Moreover, the use of local mares as

a breeding base was the general practice followed in other countries, especially Great Britain and Spain. 'Expediente sobre la Ejecución del Proyecto de Renovación y Mejora de la Cria Caballar', 1881. Such ideas also reflect the prevailing views on 'prepotency', that pedigree was supposed 'to ensure that pure-bred sires and dams exercised a disproportionate influence on resulting offspring even when mated with more ordinary animals'. Harriet Ritvo, *The Platypus and the Mermaid and other Figments of the Classifying Imagination*, Cambridge and London: Harvard University Press, 1997, p. 115.

34 'Expediente sobre Cumplimiento de la Real Orden No 275 de 30 Marzo de 1885 para el Fomento y Meyora de la Raza Caballar en las Islas, 17 September', PNA, Raza de Caballeria de Filipinas, Formento y Mejora, 1888.

35 Ibid.,'Expediente ... 9 October ...'. Another animal subsequently became ill but it is unclear whether this was as a consequence of the voyage or something contracted subsequent to arriving in the Philippines. Heavy losses at sea are blamed on the horse's inability to vomit. Wendy Doniger, 'Presidential Address: "I Have Scinde": Flogging a Dead (White Male Orientalist) Horse', *Journal of Asian Studies*, vol. 58, no. 4, 1999, p. 949.

36 'Expediente sobre Cumplimiento de la Real Orden No 275 de 30 Marzo de 1885 para el Fomento y Meyora de la Raza Caballar en las Islas, 17 September', 1888.

37 Ibid., 'Expediente ... 29 October...'; 'Expediente ... 5 November ... ', 'Expediente ... 9 November ... ', 'Expediente ... 15 November ... ', Expediente ... 7 December ...'.

38 The calculation is based on the assumption that all 25 horses were eventually sold at an average price of 643 pesos, so amounting to a total of 16,075 pesos compared to an initial outlay of 14,000 pesos. 'Expediente sobre Cumplimiento de la Real Orden No 275 de 30 Marzo de 1885 para el Fomento y Meyora de la Raza Caballar en las Islas, 5 November ...'; 'Expediente ... 9 November ...'; and 'Expediente ...15 November ...'.

39 'Memoria Preliminar para la Ejecución del Proyecto de Renovación y Mejora de la Cria Caballar', 1883.

40 De Jong Boers, 'Paardenfokkerij op Sumbawa (1500-1930)', 1997, p. 443; Doniger, 'Presidential Address: "I Have Scinde": Flogging a Dead (White Male Orientalist) Horse', 1999, p. 949; and Clarence-Smith, 'Horse Breeding in Mainland Southeast Asia and Its Borderlands', 2004. The French were also interested in improving cavalry mounts through selective crossbreeding in Algeria after 1855. General Eugène Daumas, Director of Algerian Affairs at the Ministry of War, was the author of a treatise to such effect and instrumental in the appointment of Antoine Richard du Cantal, an expert on the military and agricultural uses of large quadrupeds as President of the Permanent Commission on Algeria. Michael Osborne, *Nature, the Exotic, and the Science of French Colonialism*, Bloomington and Indianapolis: Indiana

University Press, 1994, pp. 148, 151. A similar interest in breed improvement was manifest in Lesotho; see Swart, Chapter 9, this volume.

41 Georges Buffon, *Histoire Naturelle Générale Et Particulière*. 44 vols, Paris: 1749-1804; and Jean Baptiste Lamarck, *Philosophie Zoologique*, 2 vols, Paris: [No publisher], 1809.

42 Warwick Anderson, 'Climates of Opinion: Acclimitization in Nineteenth-Century France and England', *Victorian Studies*, vol. 35, no. 2, 1992, p. 143.

43 Isidore Geoffroy Saint-Hilaire, *Acclimatation et Domestication des Animaux Utiles*, Paris: [No publisher] 1861, p. 116. Geoffroy Saint-Hilaire established the Société Zoologique et Botanique d'Acclimatation in Paris in 1854. A similar Imperial Russian Society for the Acclimatization of Animals and Plants, patterned after the French model, was formed in Moscow.

44 David Arnold, 'Tropical Medicine before Manson'. In David Arnold (ed.) *Warm Climates and Western Medicine: The Emergence of Tropical Medicine, 1500-1930*, Amsterdam and Atlanta, Georgia: Rodopi, 1996, pp. 5-10.

45 Lynn White, 'The Historical Roots of Our Ecological Crisis', *Science*, vol. 155, 1967, pp. 1203-1207.

46 Glick, 'Spain', 1974, p. 327.

47 'Memoria Preliminar para la Ejecución del Proyecto de Renovación y Mejora de la Cria Caballar', 1883.

48 Aguilar, *Colonización de Filipinas*, p. 259.

49 'Memoria Preliminar para la Ejecución del Proyecto de Renovación y Mejora de la Cria Caballar', 1883.

50 'Expediente sobre la Ejecución del Proyecto de Renovación y Mejora de la Cria Caballar', 1881; and 'Memoria Preliminar para la Ejecución del Proyecto de Renovación y Mejora de la Cria Caballar', 1883. Even among those who maintained acclimatisation was possible, some considered the sudden transference to an extreme climate as prejudicial to the health of the organism, while a more gradual exposure, through stages to different conditions, made acclimatisation to almost any part of the earth's surface possible. David Livingstone, 'Human Acclimatization: Perspectives on a Contested Field of Inquiry in Science, Medicine and Geography', *History of Science*, vol. 25, 1987, pp. 379-382.

51 Probably one of the Barrettos and a major nineteenth century entrepreneur (personal communication from Dan Doeppers, University of Wisconsin, Madison).

52 'Expediente sobre la Ejecución del Proyecto de Renovación y Mejora de la Cria Caballar', 1881.

53 Ibid.

54 Anderson, 'Climates of Opinion: Acclimatization in Nineteenth-Century France and England', 1992, p. 135.

55 'Expediente sobre la Ejecución del Proyecto de Renovación y Mejora de la Cria Caballar', 1881; and 'Memoria Preliminar para la Ejecución del Proyecto de Renovación y Mejora de la Cria Caballar', 1883.

56 'Memoria Preliminar para la Ejecución del Proyecto de Renovación y Mejora de la Cria Caballar', 1883; and *Report of the Philippine Commission to the President, January 31, 1900*, vol. 4, p. 13.

57 Spain was heavily dependent by the nineteenth century on mules from France, whose traders took advantage of the lack of local stock to export the animals across the Pyrenees 'in prodigious numbers, the French breeders carrying away in exchange a river of gold'. 'Memoria Preliminar para la Ejecución del Proyecto de Renovación y Mejora de la Cria Caballar', 1883.

58 Ibid. and E. Koert, 'Stock Breeding in the Catanduanes Islands', *Philippine Agricultural Review*, vol. 5, no. 6, 1912, p. 306. In the event, such practices were unlikely to have been practised as stallions become loud and violent when enclosed with mares and can be a danger to both them and their foals.

59 'Memoria Preliminar para la Ejecución del Proyecto de Renovación y Mejora de la Cria Caballar', 1883.

60 On the other hand, castration normally increases an animal's life-span. The parallels with feminist discourse are informative, particularly as regards gender and colonialism and concepts about Oriental or Asian 'Others'. See Vron Ware, *Beyond the Pale: White Women, Racism and History*, London and New York: Verso, 1992; Antoinette Burton, *Burdens of History: British Feminists, Indian Women, and Imperial Culture, 1865-1915*, Chapel Hill and London: University of North Carolina Press, 1994; and Anne McClintock, *Imperial Leather: Race, Gender and Sexuality in the Colonial Context*, New York and London: Routledge, 1995.

61 De Jong Boers, 'Paardenfokkerij op Sumbawa (1500-1930)', 1997, p. 443; and de Jong Boers, Chapter 4, this volume.

62 W. Best, 'Suggestions for the Care of Horses in the Philippine Islands', *Philippine Agricultural Review*, vol. 7, 1914, p. 331.

63 [No author] 'Animal Industry in the Philippines; Horses', 1911, p. 482; and H. L. Casey, Director of Studs to the Chief of the Bureau of Agriculture, Baguio, 31 August 1905. In *Report of the Philippine Commission, 1905*, Washington: Government Printing Office, 1906, p. 460.

64 Koert, 'Stock Breeding in the Catanduanes Islands', 1912, p. 307.

65 Keith Chivers, 'The Supply of Horses in Great Britain in the Nineteenth Century'. In F. Thompson (ed.), *Horses in European Economic History: A Preliminary Canter*, Reading: The British Agricultural History Society, 1983, p. 41.

66 Warwick Anderson, 'Immunities of Empire: Race, Disease, and the New Tropical Medicine, 1900-1920', *Bulletin of the History of Medicine*, vol. 70, 1996, pp. 94-118; Mark Harrison, ' "The Tender Frame of Man": Disease, Climate,

and Racial Difference in India and the West Indies, 1760-1860', *Bulletin of the History of Medicine*, vol. 70, 1996, pp. 68-93; and Philip Curtin, *Death by Migration: Europe's Encounters with the Tropical World in the Nineteenth Century*, Cambridge: Cambridge University Press, 1989, pp. 87-90.

67 Anderson, 'Climates of Opinion: Acclimatization in Nineteenth-Century France and England', 1992, p. 138.

68 Polygenism posits a separate origin for each human race and explains the subsequent diversity between races as the result of miscegenation. Monogenism, on the other hand, holds that all human races share a single primordial origin. Osborne, *Nature, the Exotic, and the Science of French Colonialism*, 1994, p. 87. As a broad generalisation, those supporting the former regarded acclimatisation as impossible and favoured economic exploitation of the tropics, while those supporting the latter believed in acclimatisation and favoured colonial settlement overseas. Livingstone, 'Human Acclimatization', 1987, p. 384.

69 Bankoff, Chapter 6, this volume.

70 Livingstone,'Human Acclimatization', 1987, p. 382; and David Livingstone, 'Tropical Climate and Moral Hygiene; The Anatomy of a Victorian Debate', *The British Journal for the History of Science*, vol. 32, 1999, p. 104.

71 'Expediente sobre la Ejecución del Proyecto de Renovación y Mejora de la Cria Caballar', 1881.

72 Curtin, *Death by Migration*, 1989, p. 42; and Warwick Anderson, '"Where Every Prospect Pleases and Only Man is Vile": Laboratory Medicine as Colonial Discourse'. In Vicente Rafael (ed.), *Discrepant Histories: Translocal Essays on Filipino Cultures*, Manila: Anvil Publishing, 1995, p. 89.

73 Lois Magner, *A History of Medicine*, New York: Marcel Dekker Inc., 1992, p. 107.

74 David Arnold, 'India's Place in the Tropical World, 1170-1930', *Journal of Imperial and Commonwealth History*, vol. 26, no. 1, 1998, p. 5.

75 R. Gonzalez y Martín, *Filipinas y Sus Habitantes*, Béjar: Establecimiento Tipográfico de la Viuda de Aguilar, 1896. pp. 21-25. José Montero y Vidal, interestingly, was one of the few Spanish authors to have a very different impression of the tropical climate: 'that of Luzon, in general, is very benign. The sky clear, fine weather and bright, the island of Luzon, like all the Philippines, is a country safe and agreeable to live in.' However, he does admit that it was better for adults than for the young. José Montero y Vidal, *El Archipiélago Filipino y las Islas Marianas, Carolinas y Palaos: Su Historia, Geografía y Estadística* [The Philippine Archipelago and the Marianas, Carolinas and Palau Islands: Their History, Geography and Statistical Profile], Madrid: Imprenta y Fundición de Manuel Tello, 1886, p. 262.

76 Felipe II, The Audiencia of Manila Re-established, 26 November 1595. In Blair and Robertson (eds), *The Philippine Islands, 1493-1898*, 1973, vol. 9, p. 190.

These instructions were meticulously carried out, using an embroidered cloth of red velvet. Pedro Desquibel, 'Reception of the Royal Seal at Manila 8 June, 1598'. In Blair and Robertson (eds), *The Philippine Islands, 1493-1898*, 1973, vol. 10, p. 135.

77 Diego de Rueda y Mendoza, 'Royal Festivities at Manila, 1 August, 1625'. In Blair and Robertson (eds), *The Philippine Islands, 1493-1898*, 1973, vol. 22, p. 59.

78 Sinibaldo de Mas, *Informe sobre el Estado de las Islas Filipinas En 1842* [Report on the State of the Philippine Islands in 1842], Madrid: Imprenta de L. Sancha, 1843, pp. 52, 62; Doniger, 'Presidential Address: "I Have Scinde": Flogging a Dead (White Male Orientalist) Horse', 1999, p. 947; Boomgaard, Chapter 3, this volume; and De Jong Boers, Chapter 4, this volume. Indio is the term used by the Spanish to denote a person indigenous to the Philippines. Mestizo refers to a person of mixed parentage, usually Chinese and Indio (more properly called a Mestizo Sangley) but also a person of mixed Spanish and Indio heritage (more properly called a Mestizo Español).

79 Gómez Pérez Dasmariñas to the King of Camboja, 27 September, 1593. In Blair and Robertson (eds), *The Philippine Islands, 1493-1898*, 1973, vol. 9, p. 78. De Jong Boers mentions that horses on Sumbawa were frequently given as 'gifts' to foreigners in exchange for opium and weaponry. De Jong Boers, 'Paardenfokkerij op Sumbawa (1500-1930)', 1997, p. 442.

80 Sebastian Hurtado de Corcuera to Felipe IV, 30 June, 1636. In Blair and Robertson (eds), *The Philippine Islands, 1493-1898*, 1973, vol. 26, p. 196; and Mas, *Informe sobre el Estado de las Islas Filipinas*, 1843: Ejercito 4. The cavalry comprised one regiment of four squadrons and six hundred mounted customs officials. The original 1843 edition of Sinibaldo de Mas's *Informe sobre el Estado de las Islas Filipinas* is not paginated consecutively throughout the work but only numbered on a section-by-section basis.

81 Alfred Crosby, *Ecological Imperialism: The Biological Expansion of Europe, 900-1900*, Cambridge: Cambridge University Press, 1993, p. 194; and Livingstone, 'Human Acclimatization', 1987, p. 380. Crosby quotes from colonial accounts of horses in the South American Pampa as existing in such numbers 'that they cover the face of the earth' or 'look like woods from a distance' in *Ecological Imperialism*, pp. 182-187.

82 Curtin, *Death by Migration*, 1989, p. 4. For Spanish records on comparable health rates among troops dating from 1884, see, Ministerio de la Guerra, *Resumen de la Estadística Sanitaria del Ejercito Española* [Summary of the Health Statistics of the Spanish Army], Madrid.

83 Anderson, '"Where Every Prospect Pleases and Only Man is Vile": Laboratory Medicine as Colonial Discourse', 1995, pp. 85-86.

Chapter 8. Riding High – Horses, Power and Settler Society in Southern Africa, c. 1654–1840
Sandra Swart

This chapter was published in an earlier incarnation as 'Riding High – horses, power and settler society, c. 1654-1840.' *Kronos*, vol. 29, Environmental History, Special Issue, Nov. 2003, and draws on themes developed first in Sandra Swart, ' "Horses! Give me more horses!" – white settler society and the role of horses in the making of early modern South Africa' in K. Raber and T. J. Tucker (eds) *The Culture of the Horse – Status, Discipline and Identity in the Early Modern World*, London: Palgrave Macmillan, 2005, pp. 311-328. I would like to record my indebtedness especially to Andrew Bank and to Bob Edgar, Malcolm Draper, Albert Grundlingh, Helena Lategan, Frans van der Merwe, Graham Walker and to the South African National Research Foundation.

1   For the horse, particularly for descriptions of how the horse played a role in major shifts in the balance of power in West Africa, see Robin Law, *The Horse in West African History*, Oxford: Oxford University Press, 1980.

2   A.W. Crosby, *Ecological Imperialism: The Biological Expansion of Europe, 900 to 1900*, Cambridge: Cambridge University Press, 1986.

3   On more general horse history see Harold Barclay, *The Role of the Horse in Man's Culture*, London: J. A. Allen & Co., 1980; Charles Chevenix-Trench, *A History of Horsemanship*, London: Doubleday, 1980; and Juliet Clutton-Brock, *Horse Power*, Cambridge: Harvard University Press, 1992. For the early South African period see P. J. Schreuder, 'The Cape Horse – its origin, breeding and development in the Union of South Africa', PhD, thesis, Cornell University, 1915; and H. A. Wyndham, *The Early History of the Thoroughbred Horse in South Africa*, London: Humphrey Milford/Oxford University Press, 1924; and J. J. Nel, 'Perdeteelt in Suid-Afrika, 1652-1752', MA thesis, University of Stellenbosch, 1930. For an analysis of the equine dimension of the period leading up to and during the South African War, see Harold Sessions, *Two Years with the Remount Commission*, London: Chapman and Hall, 1903, a useful primary source. For popular reading see Jose Burman, *To Horse and Away, Cape Town: Human and Rousseau*, 1993 or Daphne Child, *Saga of the South African Horse*, Cape Town, Howard Timmins, 1967.

4   Johan Anthoniszoon van Riebeeck (21 April 1619-18 January 1677) joined the VOC and served as an assistant surgeon in the East Indies, before undertaking the command of the initial Dutch settlement in the future South Africa, and died in Batavia on Java in 1677.

5   R. Ross, *Cape of Torments: Slavery and Resistance in South Africa*, London: Routledge, Kegan and Paul, 1983.

6   For a discussion of the nomenclature, see Shula Marks, 'Khoisan Resistance to the Dutch in the Seventeenth and Eighteenth Centuries', *Journal of Afri-*

can History, 13, 1972, pp. 55-80. 'Hottentot' referred to a pastoral grouping known as 'Khoikhoi' today and 'Bushman' to a people recognised variously today as 'Bushman' or 'San'. Today many academics refer to these pastoralists by the name they called themselves: Khoekhoen, meaning 'real people'. By contrast the 'Bushmen' were principally hunters, known as Soaqua, who had no or few domestic animals.

7   For a discussion of this see Richard Elphick, *Kraal and Castle: Khoikhoi and the Founding of White South Africa*, New Haven: Yale University Press, 1977, pp. 90-110.

8   H. C. V. Leibbrandt, *Précis of the Cape Archives: Letters Despatched 1652-62*, 2 vols, Cape Town: vol. I, 1898, vol. II, 1900, vol. I, p. 31.

9   Travelling by ox was a deeply unsatisfying mode of transport – as David Livingstone observed two centuries later. David Livingstone, *Missionary Travels and Researches in South Africa*, London: Murray, 1857.

10 For a discussion see Thomas Dunlap, *Nature and the English Diaspora – Environment and History in the United States, Canada, Australia and New Zealand*, Cambridge: Cambridge University Press, 1999.

11 Leibbrandt, *Letters Despatched* ,vol. I, 1898, no. 493, p. 670.

12 It was a rudimentary system: horses turned the ground over with a massive plough requiring eight or ten horses, and the harvested grain was trodden out by horses on circular threshing floors in the open air.

13 A. J. H. van der Walt, *Die Ausdehnung der Kolonie am Kap der Guten Hoffnung (1700-1779): eine historisch-okonomische Untersuchung uber das Werden und Wesen des Pionierlebens im 18. Jahrhunderd*, Berlin: Ebering, 1928, p. 2.

14 Leibbrandt, *Letters Despatched*, vol. I, 1898, no. 493, p. 143.

15 Ibid.

16 J. van Riebeeck, H.B. Thom (ed.), *Journal of Jan Van Riebeeck*, vol. II, 1656-1658, Cape Town: A.A. Balkema, 1954, p. 116; S. F. du Plessis, *Die verhaal van die dagverhaal van Jan van Riebeeck* [The narrative of the diary of Jan van Riebeeck], D. Litt. thesis, University of Pretoria, 1934; *Dagverhaal van Jan van Riebeek* [Diary of Jan van Riebeeck], Gravenhage: Nijhoff, 1892-1893, p 434; Leibbrandt, *Letters Despatched*, vol. I, 1898, no. 493, p. 143.

17 *Dagverhaal van Jan van Riebeeck*, 1892-1893, p. 446.

18 Leibbrandt, *Letters Despatched*, vol. I, 1898, p. 407.

19 Van Riebeeck, *Journal of Jan Van Riebeeck*, vol. I, 1954, pp. 272, 307.

20 Leibbrandt, *Letters Despatched*, vol. I, 1898, p. 1041.

21 H.C.V. Leibbrandt, *Precis of the Archives of the Cape of Good Hope: Letters Received, 1695-1708*, Cape Town: Richards, vol. I, 1898, vol. II, 1900, p. 253; and Van der Walt, *Die Ausdehnung der Kolonie*, p. 2.

22 Company servants were initially prohibited from starting local industries and freedom of trade was forbidden because free enterprise was supposed to be restricted to the VOC. Van Riebeeck argued that free burghers should be allowed to trade, farm and help with defence for some time. In 1676, VOC official policy changed: it was contended that a Dutch colony should be nurtured at the Cape to stimulate agricultural production.

23 Colin P. Groves, *Horses, Asses and Zebras in the Wild*, Sanibel: Ralph Curtis Books, 1974. The last quagga died in 1883 in an Amsterdam zoo.

24 Leibbrandt, *Letters Despatched*, no. 494, p. 943.

25 Gerrit van Spaan, *De Gelukzoeker*, Rotterdam: De Vries, 1752, p. 28; J. van Riebeeck, *Journal of Jan Van Riebeeck*, vol. III, 1954, 14 December 1660, p. 300.

26 J. van Riebeeck, *Journal of Jan Van Riebeeck*, vol. III, 1954, 14 December 1660, p. 300.

27 Peter Kolbe, *Naaukeurige en uitvoerige beschryving van Kaap de Goede Hoop*, Amsterdam: B. Lakeman, 1727, p. 73.

28 This does not appear to have been the case without exception. Thomas Baines, for example, recorded that 'even the wild quaggas were dying of the horse-sickness', in March 1850, *Journal of Residence in Africa*, vol. II, Cape Town, The Van Riebeeck Society, 1961, p. 217.

29 A. Sparrman, *A Voyage to the Cape of Good Hope, Towards the Antarctic Polar Circle, and Around the World, But Chiefly into the Country of the Hottentots and Caffres, From the Year 1772 to 1776* [1785], Cape Town: Van Riebeeck Society, 1975-1977, pp. 216-217.

30 A span of zebras were even briefly utilised to run stagecoaches to Pietersburg at end of nineteenth century.

31 Captain Horace Hayes broke a young Burchell's zebra in Pretoria in 1892. The Mazawattee Company also ran two four-in-hand teams of zebras for advertising purposes. See Clive Richardson, *The Horse Breakers*, London: J.A. Allen, 1998, pp. 121-122. A professor of natural history at Edinburgh from 1882-1927, Cossar Ewart crossed a zebra stallion with pony mares in order to disprove telegony, and a secondary aim of these experiments was to produce a draught animal for South Africa that was less subject to local diseases. W.J. Broderip, *Zoological Recreations*, London: H. Colburn, 1847; Special Collections L.14.29, University of Edinburgh.

32 A 'hand' measures 4 inches or 10.16 cm.

33 Schreuder, 'The Cape Horse', 1915.

34 Later a handful of these were recaptured (1655 and 1656) and brought to the Cape.

35 Sumbawa entered into treaty relations with the Dutch East India Company in 1674, but supplied horses earlier and established itself as chief horse purveyor.

36 Leibbrandt, *Letters Despatched*, vol. I, 1898 no. 493, p. 1118.

37 Leibbrandt, *Letters Despatched*, vol. II, 1900, p. 311; *Letters Despatched*, vol. I, 1898 no. 493, pp. 1098-99. Similarly, in West Africa, for example, horses were often used as symbolic demonstrations of power, in contexts where they were of little military value. Law, *The Horse in West African History*, p. 193.

38 Leibbrandt, *Letters Despatched*, vol. II, 1900 no. 494, p. 65.

39 Leibbrandt, *Letters Despatched*, vol. I, 1898 no. 493, p. 1098. Later, in the early nineteenth century, when the San suffered massive suppression, the traveller Burchell recorded that 'A troop of horsemen is the most alarming sight which can present itself to a kraal [homestead] of Bushmen [San] in an open plain, as they then give themselves up for lost, knowing that under such circumstances, there is no escaping from these animals.' Burchell, *Travels in the Interior of Southern Africa*, 1824, p.187.

40 B. Tuchman, *The Proud Tower*, Macmillan, New York, 1966. This is buttressed by psycho-social analysis by, for example, Freud, who described the horse as symbol of power. Psychologists suggest that literally 'looking down on others' from the back of a horse may increase feelings of pride and self-esteem. D. Toth, 'The Psychology of Women and Horses'. In GaWaNí PonyBoy (ed.), *Of Women and Horses*, Irvine: BowTie, 2000, p. 36.

41 A horse at pasture requires an acre to sustain its nutritional needs, as a rule, although this varies according to the type of horse, energy output of horse, the nature of the grass, and seasonal climatic variation.

42 As observed in a 1603 text, 'in order to sustain political societies, and in order to protect the people who make up these societies, you cannot do without the horse'. Quoted in Pia Cuneo, 'Beauty and the Beast: Art and Science in Early Modern European Equine Imagery', *Journal of Early Modern History*, vol. 4, nos. 3-4, 2000, p. 269. Seventeenth century European iconographic representations of the horse emphasised serenely regal control by (patrician) riders.

43 Lisa Jardine and Jerry Brotton, *Global Interests: Renaissance Art between East and West*, Ithaca: Cornell University Press, 2000, p. 145.

44 L. Thompson, *A History of South Africa*, New Haven, Yale University Press, 1990, p. 35.

45 In any event, the various differentiated breeds of horses were simply not available.

46 Horse maiming was sometimes undoubtedly a form of social rebellion in Europe, for example. It was also a form of psychological terror, of symbolic

murder, that resulted from personal feuds between members of the same social class. Roger Yates, Chris Powell, Piers Beirne, 'Horse Maiming in the English Countryside: Moral Panic, Human Deviance, and the Social Construction of Victimhood', *Society & Animals*, vol. 9, no. 1, 2001, pp. 1–23. G. Elder, J. Wolch, and J. Emel, 'Race, place, and the bounds of humanity', *Society & Animals*, vol. 6, no. 2, 1998, pp. 183–202.

47 Elder, Wolch and Emel, 'Race, place, and the bounds of humanity', p. 184.

48 For a discussion of Cape defences see O. F. Mentzel, *Life at the Cape in the Mid-eighteenth Century* [1784], trans. M. Greenless, reprinted Cape Town: Darter Brothers, 1920, pp. 146–156.

49 Members of the kommando were expected to provide their own mount and saddlery, rifle and 30 rounds of ammunition. The *kommando* was made up of mounted marksmen, without uniforms and formal training. Christiaan Grimbeek, 'Die Totstandkoming van die Unieverdegingsmag met Spesifieke Verwysing na die Verdedigingswette van 1912 en 1922' [The Establishment of the Union Defence Force with Specific Reference to the Defence Acts of 1912 and 1922], D.Phil thesis, University of Pretoria, 1985, p 7.

50 P. H. Frankel, *Pretoria's Praetorians, Civil-military Relations in South Africa*, Cambridge: Cambridge University Press, 1984.

51 Law, *The Horse in West African History*, 1980, p. 184.

52 Wyndam, *History of the Thoroughbred*, 1924, p. 97. Moreover, from the time the Sotho, for example, overcame socio-political and economic hurdles and acquired the first horse, the 'ox without horns' as they called it, in 1825, they re-invented themselves as an equine society. Just as in the Americas, horses endowed their owners with enhanced military capabilities, as well as hunting and transport capacity. Greater mobility in turn meant greater involvement in trade networks. Eric Wolf, *Europe and the People Without History*, Berkeley: University of California Press, 1982, 1997, p. 177. For a discussion of the equine revolution in Sotho society, for example, see Sandra Swart, 'The "Ox That Deceives": The Meanings of the "Basotho Pony" in Southern Africa', Chapter 9, this volume.

53 See the *Cape of Good Hope Government Gazette*, 17 September 1830, no. 1288.

54 It does not spread directly between horses, but is transmitted by midges, which become infected when feeding on infected horses. It occurs mostly in the rainy season when midges are abundant, and fades after the frost, when the midges die. Horses become infected between sunset to sunrise, when the midges are active.

55 *Origineel Plakkaat Boek*, 1735, 1753, p. 82.

56 Child, *The Saga of the South African Horse*, 1967, p. 13.

57 Ibid., p. 22.

58 George McCall Theal, *History and Ethnography of Africa South of the Zambesi*, vol. II, London: George Allen and Unwin, 1907, 1922, p. 323; Leibbrandt, *Letters Received*, no. 417, p. 618.

59 These animals are believed to have been of the early English Roadster breed. There is evidence to show that the name Roadster was synonymous with Hackney. M.H. Hayes, *Points of the Horse*, New York: Arco, 1969.

60 The tamed mounts transported by the Spanish to the Americas assumed a feral state in the New World and developed into the 'mustang'.

61 See, for example, Elizabeth Lawrence, 'Rodeo Horses: the wild and the tame'. In R. Willis, *Signifying Animals – Human Meaning in the Natural World*, London: Routledge, 1990, p. 223 and more extensively Elizabeth Lawrence, *Rodeo: An Anthropologist Looks at the Wild and the Tame*, Knoxville: University of Tennessee Press, 1981. See also Richard Slatta, *Cowboys of the Americas*, New Haven: Yale University Press, 1990.

62 The *tripple* is further discussed later in this chapter. The trot, in contrast, is a two-beat jarring ride, with legs moving in diagonal pairs, which is very useful for pulling vehicles and horses that trot were better at racing and jumping than the gaited types. Gaited sorts had been prized prior to the seventeenth century for general purpose riding because of their ability to cover much ground in comfort. So it was wheeled travel on well-built roads and the rising popularity of both jumping and racing that marginalised the gaited horses in western Europe. See Karen Raber and Treva J. Tucker (eds), *The Culture of the Horse: Status, Discipline, and Identity in the Early Modern World*, New York: Palgrave Macmillan, 2005.

63 The first hundred years saw no major new breed importations, resulting in a foundation stock of hardy ponies.

64 John Barrow, *Travels in the Interior of South Africa, in the Years 1797 and 1798*, New York, N.Y.: Johnson Reprint, 1968, p. 36.

65 The horses of Spanish origin were taken from two French ships boarded during the Napoleonic wars.

66 From 1769, the Cape Horse gained escalating renown as an army remount and was exported to India for use by the British over the next century.

67 W. Bird, *State of the Cape of Good Hope in 1822*, reprinted Cape Town: Struik, 1966, p. 99.

68 F. J. van der Merwe, *Ken Ons Perderasse*, Cape Town: Human and Rousseau, 1981.

69 George Thompson, *Travels and Adventures in Southern Africa*, London: Henry Colburn, 1827.

70 H. Lichtenstein, *Travels in Southern Africa in the Years 1803, 1804, 1805 and 1806*, Cape Town: The Van Riebeeck Society, 1928–1930. See also Anon,

Sketches of India, or, Observations Descriptive of the Scenery, etc. in Begal ... Together with Notes on the Cape of Good-Hope, and St. Helena, London: Black, Parbury and Allen, 1816, p. 74.

71 R. Ross, Beyond the Pale, Essays on the History of Colonial South Africa, Hanover: Wesleyan University Press, 1993, p. 1; L. Thompson, A History of South Africa, p. 41.

72 M. H. Hayes, *Among Horses in South Africa*, London: Everett, 1900. Lichtenstein still makes mention of the trot as a gait used by trekboers, or semi-nomadic farmers, in the late eighteenth century.

73 This 'peculiar pace' was mentioned in 1806, but was labelled 'the bungher' (this could be a corruption of 'burgher'), and contended that an English jockey soon '[got] rid of it'. Anon. *Gleanings in Africa, Exhibiting a Faithful and Correct View of the Manners and Customs of the Inhabitants of the Cape of Good Hope, and Surrounding Country*, 1806, New York: Negro Universities Press, 1969, p.23.

74 Lady Duff Gordon, *Letters from the Cape*, Cape Town: Maskew Miller, 1925, p. 84.

75 See the first chapter of H. Rider Haggard, *Jess* [1887], London: Smith, Elder, 1909.

76 See Chapter IV, 'Doctor Rodd', in H. Rider Haggard, *Finished* [1917], London: Macdonald, 1962. This later became something that was cherished as traditional Boer riding style; see George Green, 'How to train a horse to triple', *Farmers Weekly*, vol. 74, no. 55, 24 December 1947.

77 George McCall Theal, *Records of the Cape Colony*, London: Clowes, 1897–1905, vol. XIV. p. 23.

78 See Letter to Huskisson, *Records*, vol. III, p. 390. See Barrow, *Travels in the Interior of South Africa*, vol. I, p. 67.

79 Schreuder, 'The Cape Horse', 1915, p. 39. In England a horse's pedigree had become perceived as vital far earlier; in 1791 a General Stud Book was introduced.

80 *The Agricultural News and Farmer's Journal*, 7 February 1850.

81 The state attempted breed improvement through the importation of Hackneys from Britain in 1888 and 1891.

82 Anon, *Sketches of India*, 1816, p. 75.

83 Charles Thunberg, *Travels in Europe, Africa and Asia, Performed in the Years 1770 and 1779*, vol. II, London, 1793. Law notes a similar phenomenon in West Africa, where mares had to feed themselves by grazing and mares were unpopular for riding. *The Horse in West African History*, p. 45. The parallel was extended to human society: while Boer women were not expected to be equestrians, the South African war hero, General C. De Wet, famously maintained that a 'Boer without a horse is only half a man'.

84 Wyndham, *History of the Thoroughbred*, 1924, p. 2.

85 The initial judgment appears to have been that he was an interfering idiot who had lost his life unnecessarily. In the first report to Holland his name was not even mentioned, though considerable space was devoted to the 18 boxes of money that had been salvaged. When Woltemade's body washed ashore on the 2 June 1773, he was buried without ceremony in an unmarked grave. Later, however, a VOC ship was named after him, his widow received compensation, and in 1956 a statue was erected in his honour, after the idea was mooted in the Van Riebeeck festival in 1952, and a commemorative stamp was issued in 1973. See Sparrman, *A Voyage to the Cape of Good Hope*, pp. 126–127.

86 Those like Dick Turpin's 'Black Bess' appear to be later inventions teleologically imposed on the early eighteenth century.

87 King sought to alert the British about the need for reinforcements in a campaign against the Boers (Afrikaners) in Natal in 1842. Until recently the name of his fellow rider Ndongeni Ka Xoki was seldom mentioned and the name of his horse is unknown. An equestrian monument to Dick King on the Victoria embankment in Durban was erected in 1915.

88 Like Boer General C. De Wet's famous grey 'Fleur' in the South African War (1899–1902).

89 Interestingly, recent research has shown that Napoleon's famous horse, Marengo, may be a creation of the public imagination. Official and unofficial representations of the fiery little Arab stallion, by name or presumption Marengo, came from several different horses. Equally revealing of the social significance of horsemanship as signifier of power was that Napoleon was a mediocre rider, and his frequent involuntary descents were hidden from the public. See Jill Hamilton, *Marengo: The Myth of Napoleon's Horse*, London: Fourth Estate, 2000.

90 Raber contends that the exotic 'orientalised' Other, to use Said's categorisation, was nationalised: the Arabian, Turk and Barb modified into the English Thoroughbred. Moreover, the continental *haute école* was replaced by specifically British styles. As Raber notes '[w]hile the new horse culture might still be socially restrictive, as a metaphor and interpretive device it became an inclusive definition of Englishness.' Raber and Tucker (eds), *Culture of the Horse*, 2005, p. 36.

91 Wyndham, *History of the Thoroughbred*, 1924, p. 16.

92 This blood stock formed the foundation of the Melck stud, which from c. 1808 to 1881 was an influential stud in southern Africa.

93 W. H.C. Lichtenstein, *Foundation of the Cape, About the Becuanas: Being a History of the Discovery and Colonisation of South Africa* [1811, 1807], trans. O. H. Spohr, Cape Town: Balkema, 1973.

94 Anon. *Gleanings in Africa* [1806], 1969, p. 23.

95 William Burchell, *Travels in the Interior of Southern Africa*, vol. II, London, Batchworth, 1822, reprinted 1953, p. 25.

96 Duff Gordon, *Letters from the Cape*, pp. 152-153. She also observed that the Malays maintained a grip on the horse-hiring industry, p. 46.

97 Anon. *Gleanings in Africa* [1806], 1969, p. 23. Burchell, *Travels in the Interior of Southern Africa*, vol. II, p. 163.

98 Lady Anne Barnard, *The Cape Diaries of Lady Anne Barnard, 1799-1800*, vol. I, M. Lenta and B. le Cordeur (eds), Cape Town: The Van Riebeeck Society, 1999. pp. 311, 333-334 and 359. Lady Anne Barnard, *The Letters of Lady Anne Barnard to Henry Dundas, from the Cape and Elsewhere, 1793-1803, Together with her Journal of a Tour into the Interior, and Certain Other Letters*, A. M. Lewin Robinson (ed.), Cape Town: Balkema, 1973, September 1798. Interestingly, she also mentions oxen or bullock racing among Khoikhoi, pp. 116, 123.

99 Wyndham, *History of Thoroughbred*, 1924, p. 25.

100 The British returned the Cape Colony to the Netherlands in 1803.

101 Wyndham, *History of the Thoroughbred*, 1924, p.34.

102 Ibid., p. 39.

103 Cowper Rose, *Four Years in Southern Africa*, London: Colburn and Bentley, pp. 9-10.

104 Bird, *State of the Cape of Good Hope in 1822*. He mentions also the availability of mules, chiefly from Buenos Aires, Brazil, at 200 *rixdollars*, and Spanish jacks selling at 500-1,000 *rixdollars*, p. 99.

105 Burchell, *Travels in the Interior of Southern Africa*, vol. I, 1824, p. 25.

106 Bird, *State of the Cape of Good Hope in 1822*, p. 163.

107 Wyndham, *History of the Thoroughbred*, 1924, p.22.

108 *Gazette*, August 1814.

109 Ibid.

110 A. E. Blount, *Notes on the Cape of Good Hope Made During an Excursion in that Colony in the Year 1820*, London: Murray, 1821.

111 The first advertisement for a stallion at stud appears to have been 25 September 1805, in the *Gazette*.

112 KAB, GH, NO.1/46, 1825; KAB, GH, NO.1/58, 1826.

113 Wyndham, *History of the Thoroughbred*, 1924, pp. 88-89.

114 Signed C:B: Esqr delt; pubd by Dighton, Spring Gardens, Decr. 1811.

Chapter 9. The 'Ox That Deceives': The Meanings of the 'Basotho Pony' in Southern Africa

Sandra Swart

My thanks to my companion in the field, Graham Walker, and the participants at the International Conference on forest and environmental history of the British

Empire and Commonwealth, University of Sussex, March 2003. Also to William Beinart, Robert Edgar, Richard Grove, Albert Grundlingh, Natasha Distiller, Frans van der Merwe and Lance van Sittert. Elements of this chapter appeared as ' "Race" horses – a discussion of horses and social dynamics in post-Apartheid Southern Africa' in N. Distiller and M. Steyn (eds), *Under Construction: 'Race'and Identity in South Africa Today,* Heinemann, 2004, pp. 13–24. I have used SeSotho (instead of South African) orthography: SeSotho refers to language and customs; MoSotho is a person; BaSotho refers to people, Lesotho refers to the country and LeSotho refers to the nation. I have used the term 'Basotho pony', but it is often commonly referred to as 'Basuto pony'.

1   More than 80 per cent of the country is 1,800 metres above sea level. Basuto-land was renamed the Kingdom of Lesotho upon independence from the UK in 1966. King Moshoeshoe II was in exile but was reinstated in 1995, and constitutional government was restored in 1993 after 23 years of military rule.

2   ANC Daily News Briefing, Thursday 13 July 1995 @ MANDELA-HORSE MASERU, www.anc.org.za/anc/newsbrief/1995/news0713. The Lesotho monarch, who owns a number of race horses, presented Mandela with the gift from the royal stable. The gift was made at the royal palace in Maseru after Mandela, Lesotho Prime Minister Ntsu Mokhehle and the king met. Accepting the gift, Mandela said: 'This reminds me of my younger days when I was a shepherd.'

3   P. Savory, *Basuto Fireside Tales.* Cape Town: Howard Timmins, 1962. Horses are understandably absent from the older myths and legends, see E. Jacottet, *The Treasury of Ba-Suto Lore.* Morija: Sesuto Book Depot, 1908; and M. Postma, *Tales from the BaSotho*, Trans. S. McDermid. Austin: University of Texas Press, 1974.

4   There is no historical or contemporary role for the horse among women; it was and remains strictly a male preserve. In K. Limakatso Kendall's *Basali! Stories by and about women in Lesotho*, Pietermaritzburg: University of Natal Press, 1995, other animals, but not horses, are mentioned. Gendered horse ownership in Lesotho offers a useful counter narrative to the current feminisation of equestrianism, currently the dominant discourse in the west, particularly Anglophone countries.

5   Grain and hides, for example, were traded for iron from the Transvaal area.

6   About 30 kilometres (20 miles) from what is now the capital, Maseru.

7   G. Tylden, *The Rise of the Basuto*, Cape Town: Juta, 1950, p. 9. See Sandra Swart, Chapter 8, this volume.

8   Ibid.

9 E. Eldridge, *A South African Kingdom – The Pursuit of Security in Nineteenth Century Lesotho*, Cambridge: Cambridge University Press, 1993, p. 26; D. F. Ellenberger, *History of the Basuto*, New York: Negro University Press, 1969, p. 195; Peter Sanders, *Moshoeshoe: Chief of the Sotho*, London: Heinemann, 1975, p. 46. See also Russel William Thornton, *The Origin and History of the Basotho Pony*, Morija, Morija Printing Works, 1936. There is some dispute over when and how King Moshoeshoe himself acquired his first horse. Ellenberger suggests it was sent to him by a neighbouring leader; there is anecdotal suggestion that the first horses were brought to the area by the Prussian botanist Zydensteicker (Seidenstecher) in 1829 or 1830. Ellenberger, *History of the Basuto*, p. 195; Sanders, *Moshoeshoe*, p. 46.

10 See Kolbe, 'The present state of the Cape of Good Hope, 1713', trans. G. Medley, London: W. Innys and R. Manly, p. 33; *Cape Quarterly Review*, January 1882, p. 273, 'khomohaka', the deceiver ox (without horns).

11 Ellenberger, *History of the Basuto*, p. 195.

12 William Lye (ed.), *Andrew Smith's Journal of his Expedition into the Interior of South Africa, 1834-36*, Cape Town: A. A. Balkema, 1975, October 1834, p. 65.

13 Tylden, *The Rise of the Basuto*, p. 9.

14 Lye (ed.), *Andrew Smith's Journal of his Expedition into the Interior of South Africa, 1834-36*, October 1834, p. 64.

15 Numbers drawn from Eldridge, *A South African Kingdom*, 1993, p. 147.

16 Minnie Martin, *Basutoland – Its Legends and Customs*, London: Nichols, 1903, p. 40.

17 Tylden, *The Rise of the Basuto*, p. 118.

18 Eldridge, *A South African Kingdom*, p. 160.

19 B.C. Judd, 'With the C. M. R', *Nongquai*, June 1938.

20 As in the Philippines, see Greg Bankoff, Chapter 6, this volume.

21 J. M. Christy, *Transvaal Agricultural Journal*, 1908; P. J. Schreuder, 'The Cape Horse', PhD thesis, Ithaca: Cornell University, 1915, p. 101.

22 Tylden, *The Rise of the Basuto*, p. 208.

23 National Archives, Public Record Office, London, DO119/1096.

24 National Archives, Public Record Office, London, DO119/1096.

25 Memorandum of Understanding Concerning the Basotho Pony Project, 1 July 1977, Maseru. Lesotho Treaty List, no.5 of 1980, p. 22; on 8 May 1973, during a donor conference, Ireland agreed to establish the Basotho Pony Project to respond to the demand from countries where the Basotho Pony's potential was known. In 1978, the National Stud was established under an agreement reached between the Governments of Lesotho and Ireland in 1976.

26 McCormack, 1991, p. 201.

27 H. Sessions, *Two Years With Remount Commission*, London: Chapman and Hall, 1903, p. 215.

28 J. L. Lush, *The Genetics of Populations*, Mimeo: University of Iowa Press, 1948; see 'From Jay L. Lush to Genomics: Visions for Animal Breeding and Genetics', Conference Proceedings, May 16–18, 1999, Department of Animal Science, Iowa State University.

29 There are about 150 internationally accepted 'breeds' of horses in the world, all belonging to the species, *Equus caballus*.

30 Although anthropologists no longer classify populations in terms of races, they do recognise that human populations exhibit diverse phenotypes. Ancestral phenotypes are suites of traits that are associated with geographic populations.

31 'Conservation of Early Domesticated Animals of Southern Africa', Conference proceedings, Willem Prinsloo Agricultural Museum, 3–4 March 1994.

32 F. van der Merwe and J. Martin, unpublished paper [n.d]. My thanks for a copy of this paper.

33 www.orusovo.com/weissenfels/nooitgedacht.htm.

34 www.nooitgedacht.com/nootie_heritage.htm.

35 Nooitgedacht Horse Breeders' Society, www.sa-breeders.co.za/org/nooitgedachter/

36 Ania Loomba, *Colonialism/Postcolonialism*, London and New York: Routledge, 1998.

37 Robert Young, *Colonial Desire: Hybridity in Theory, Culture and Race*, London: Routledge, 1995, p. 10.

38 Loomba, *Colonialism/Postcolonialism*, p. 173.

39 Animals are used as signifiers in an attempt to boost post-colonial pride in indigenous identity. This is not limited to the equine sphere. The singing dogs of New Guinea, Korean Jindo or Australian dingo (*Canis familiaris dingo*), for example, are increasingly argued to be 'breeds' in their own right. The dingo, for example, is currently celebrated as 'part of the living heritage of aboriginal culture and part of Australian history'. The Jindo of Korea has been described recently as 'one of the Korean natural monuments', around 'from time unknown'. C.G. Lee et al, 'A Review of the Jindo, Korean Native Dog', *Asian-Australasian Journal of Animal Sciences*, vol. 13, no. 3, 2000, pp. 381–389.

40 Simon Dalby 'Ecological Politics, Violence, and the Theme of Empire', *Global Environmental Politics*, vol. 4, no. 2, May 2004, pp. 1–11.

Chapter 10. 'Together yet Apart': Towards a Horse-story
Greg Bankoff and Sandra Swart

1 Lynn White, *Medieval Technology and Social Change*, Oxford: Oxford University Press, 1962. White argues that the invention of the horse-stirrup gave rise to a warrior class around which the structure of feudal society arose.

2 Partha Chatterjee, *The Nation and Its Fragments: Colonial and Postcolonial Histories*, Princeton, New Jersey: Princeton University Press, 1993, p. 224.

3   On an innovative approach towards animal history, see, in particular, Jason Hribal '"Animals are Part of the Working Class": A Challenge to Labor History', *Labor History*, vol. 44, no. 4, 2003, pp. 435-453.

4   Aldo Leopold, *A Sand County Almanac with Essays on Conservation from Round River*, New York: Ballantine Books, 1970, pp. 239-246. Leopold refers to a *land ethic* where the notion that all individuals must be members of a community of interdependent parts is simply enlarged to include the soils, waters, plants as well as the animals.

Bibliography

Adriani, Nicolaus (1928) *Bare'e-Nederlandsch Woordenboek* [Bare'e-Dutch dictionary], Leiden: Brill [Bataviaasch Genootschap van Kunsten en wetenschappen].

Aguilar, Filomeno (1998) *Clash of Spirits: The History of Power and Sugar Planter Hegemony on a Visayan Island*, Honolulu: University of Hawai'i Press.

Aguilar, José. (1893) *Colonización de Filipinas* [The Colonisation of the Philippines], Madrid: Establecimiento Tipográfico de Alfredo Alonso.

Ajoebar (1927) 'De Verbetering van den Paardenstapel op West-Soembawa' [The Improvement of the Horse-Population in Western-Sumbawa], *Koloniale Studiën*, vol. 11 (part II) (pp. 366-373).

ANC Daily News Briefing, Thursday 13 July 1995 @ MANDELA-HORSE MASERU, www.anc.org.za/anc/newsbrief/1995/news0713.

Anderson, John (1971) *Acheen, and the Ports on the North and East Coasts of Sumatra*, Oxford University Press.

Anderson, Kay (1997) 'A Walk on the Wild Side: A Critical Geography of Domestication', *Progress in Human Geography*, vol. 21 (pp. 463-485).

Anderson, Virginia DeJohn (2004) *Creatures of Empire: How Domestic Animals Transformed Early America*, New York: Oxford University Press.

Anderson, Warwick (1992) 'Climates of Opinion: Acclimatization in Nineteenth-Century France and England', *Victorian Studies*, vol. 35, no. 2 (pp. 135-157).

—— (1995) '"Where Every Prospect Pleases and Only Man is Vile": Laboratory Medicine as Colonial Discourse'. In Vicente Rafael (ed.), *Discrepant Histories: Translocal Essays on Filipino Cultures*, Manila: Anvil Publishing (pp. 83-112).

—— (1996) 'Immunities of Empire: Race, Disease, and the New Tropical Medicine, 1900-1920', *Bulletin of the History of Medicine*, vol. 70 (pp. 94-118).

'Animal Industry in the Philippines; Horses' (1911) *Philippine Agricultural Review*, vol. 4, no. 9 (pp. 476-483).

Anon (1816) *Sketches of India, or, Observations Descriptive of the Scenery, etc. in Begal ... Together with Notes on the Cape of Good-Hope, and St. Helena*, London: Black, Parbury and Allen.

Anon (1969) *Gleanings in Africa, Exhibiting a Faithful and Correct View of the Manners and Customs of the Inhabitants of the Cape of Good Hope, and Surrounding Country, 1806*, New York: Negro Universities Press.

Arkin, Marcus (1973) *Storm in a Teacup, the Later Years of John Company at the Cape, 1815-36*, Cape Town: Struik.

Arnold, David (1996) 'Tropical Medicine before Manson'. In David Arnold (ed.) *Warm Climates and Western Medicine: The Emergence of Tropical Medicine, 1500-1930*, Amsterdam and Atlanta, Georgia: Rodopi (pp. 1-19).

—— (1998) 'India's Place in the Tropical World, 1170-1930', *Journal of Imperial and Commonwealth History*, vol. 26, no. 1 (pp. 1-21).

Avellano, Manuel (1912) *Geografía de las Islas Filipinas*, Manila: Tipografía del Colegio de Sto. Tomás.

Baines, Thomas (1961) *Journal of Residence in Africa*, vol. II, Cape Town: The Van Riebeeck Society.

Baker, Steve (1993) *Picturing the Beast: Animals, Identity and Representation*, Manchester: Manchester University Press.

Balfour, Edward (1871) *Cyclopaedia of India and of Eastern and Southern Asia, Commercial, Industrial and Scientific*, Madras.

Ball, Caroline (1994) *Horse and Pony Breeds*, Edison: Chartwell.

Ballot, J. (1897) 'Historisch Overzicht van de Maatregelen tot Verbetering van den Vaarden- en Veerstapel in Nederlandsch-Indië' [Historical Review of Measures to Improve the Horse and Cattle Population in the Netherlands-Indies], *Veeartsenijkundige Bladen voor Nederlandsch-Indië*, vol. 10 (pp. 21-87).

Bankoff, Greg (1991) 'Redefining Criminality: Gambling and Financial Expediency in the Colonial Philippines 1764-1898', *Journal of Southeast Asian Studies*, vol. 22 (pp. 267-281).

—— (2001) 'A Question of Breeding: Zootechny and Colonial Attitudes Towards the Tropical Environment in Late Nineteenth Century Philippines', *Journal of Asian Studies*, vol. 60, no. 2 (pp. 413-437).

—— (2003) *Cultures of Disaster: Society and Natural Hazard in the Philippines*, Richmond: Routledge Curzon.

—— (2004a) '*Bestia Incognita*: The Horse and Its History in the Philippines 1880-1930', *Anthrozoös*, vol. 17, no. 1 (pp. 3-25).

—— (2004b) 'Horsing Around: The Life and Times of the Horse in the Philippines at the Turn of the 20th Century'. In Peter Boomgaard and David Henley (eds), *Smallholders and Stockbreeders: Histories of Foodcrop and Livestock Farming in Southeast Asia*, Leiden: KITLV Press, (pp. 233-255).

Barclay, Harold B. (1980) *The Role of the Horse in Man's Culture*, London: J. A. Allen & Co.

Barendse, R. J. (2001) 'Reflections on the Arabian Seas in the Eighteenth Century', *Itinerario*, vol. 25, no. 1 (pp. 25-49).

Barnard, Lady Anne (1973) *The Letters of Lady Anne Barnard to Henry Dundas, from the Cape and Elsewhere, 1793-1803, Together with her Journal of a Tour into the Interior, and Certain Other Letters*, A. M. Lewin Robinson (ed.), Cape Town: Balkema.

—— (1999) *The Cape Diaries of Lady Anne Barnard, 1799-1800*, vol. I, M. Lenta and B. le Cordeur (eds), Cape Town: The Van Riebeeck Society.

Barrow, John (1968) *Travels in the Interior of South Africa, in the Years 1797 and 1798*, New York: N.Y. Johnson, Reprint 1968.

Bayly, C. A. (1990) *Indian Society and the Making of the British Empire*, Cambridge: Cambridge University Press.

Bennett, Charles (1960) 'Cultural Animal Geography: An Inviting Field of Research', *Professional Geographer*, vol. 12, no. 5 (pp. 12-14).

Berg, L.W.C. van den (1886) *Le Hadhramout et les Colonies Arabes dans l'Archipel Indien*, Batavia: Imprimerie du Gouvernement.

Berger, J. (1971) 'Animal World', *New Society* (p. 1043).

Berger, John (1980) 'Why look at animals?' In John Berger, *About Looking*, New York: Pantheon Books.

Best, W. (1914) 'Suggestions for the Care of Horses in the Philippine Islands', *Philippine Agricultural Review*, vol. 7 (pp. 327-331).

Bird, W. (1966) *State of the Cape of Good Hope in 1822*, reprinted Cape Town: Struik.

Birkhead, Tim (2003) *A Brand-New Bird: How Two Amateur Scientists Created the First Genetically Engineered Animal*, New York: Basic Books.

Blake, George (1956) *B. I. Centenary, 1856-1956*, London: Collins.

Blas Jerez to Director-General de Administración Civil (1889) Philippine National Archives, Carruajes, Carros y Caballos, Bundle 7.

Blok, Gisela (1991) 'Challenging Dichotomies: Perspectives on Women's History'. In Karen Offen, Ruth Roach Pierson and Jane Rendall (eds), *Writing Women's History, International Perspectives*, Basingstoke, Hampshire: Macmillan (pp. 1-23).

Blount, A .E. (1821) *Notes on the Cape of Good Hope Made During an Excursion in that Colony in the Year 1820*, London: Murray.

Boer, Michael Georg de and Johannes Cornelis Westermann (1941) *Een Halve Eeuw Paketvaart, 1891-1941* [Half a Century of Packet-Shipping], Amsterdam: DeBussy.

Bogdan, Robert and Steven J. Taylor (1975) *Introduction to Qualitative Research Methods*, New York: John Wiley & Sons.

Bongianni, Maurizio (ed.) (1988) *Simon & Schuster's Guide to Horses and Ponies*, New York: Simon and Schuster.

Boomgaard, Peter (1999) 'Maize and Tobacco in Upland Indonesia'. In Tania Murray Li (ed.), *Transforming the Indonesian Uplands: Marginality, Power and Production,* Amsterdam: Harwood Academic Publishers.

—— (2001) *Frontiers of Fear: Tigers and People in the Malay World 1600-1950*, New Haven: Yale University Press.

—— (2004) 'Horses, Horse Trading and Royal Courts in Indonesian History, 1500-1900'. In Peter Boomgaard and David Henley (eds), *Smallholders and*

Stockbreeders: History of Foodcrop and Livestock Farming in Southeast Asia, Leiden: KITLV Press. [Verhandelingen 218] (pp. 211-232).

Boomgaard, Peter and David Henley (eds) (2004) *Smallholders and Stockbreeders: Histories of Foodcrop and Livestock Farming in Southeast Asia*, Leiden: KITLV Press.

Bouman, M.A. (1925) 'Toeharlanti: De Bimaneesche Sultansverheffing' [Tuharlanti: The Bimanese Installation of a Sultan], *Koloniaal Tijdschrift* 14 (pp. 710-717).

Braam Morris, D.F. van (1891) 'Nota van Toelichting bij het Contract Gesloten met het Landschap Bima 1886' [Explanatory Memorandum to the Contract Concluded with the Region of Bima 1886], *Tijdschrift voor Indische Taal-, Land- en Volkenkunde (TBG)* 34 (pp. 176-233).

Brenner, J. Freiherr von (1894) *Besucht bei den Kannibalen Sumatras: Erste Durchquerung der Unabhängigen Batak-Lande* [Visit to the Cannibals of Sumatra: First Cut Right Across the Independent Batak Area], Würzburg: Woerl.

British Burma Gazette (1880) Rangoon: Government Press.

British South Africa Company (1900) *Reports on the Administration of Rhodesia, 1898-1900*, London.

Broderip, W.J. (1847) *Zoological Recreations*, London: H. Colburn, 1847; Special Collections L.14.29, University of Edinburgh.

Broeze, Frank J. A. (1979) 'The Merchant Fleet of Java, 1820-1850: A Preliminary Survey', *Archipel*, vol. 18 (pp. 251-69).

Brummelhuis, Han ten (1987) *Merchant, Courtier and Diplomat: A History of the Contacts between the Netherlands and Thailand*, Lochem-Gent: De Tijdstroom.

Buffon, Georges (1794-1804) *Histoire Naturelle Générale Et Particulière*. 44 vols, Paris.

Bulliet, Richard (1975) *The Camel and the Wheel*, Cambridge: Harvard University Press.

Burchell, William (1822, 1953) *Travels in the Interior of Southern Africa*, vol. II, London: Batchworth.

Burman, Jose (1993) *To Horse and Away*, Cape Town: Human and Rousseau.

Burrill, Harry R. and Raymond F. Crist (1980) *Report on Trade Conditions in China*, New York: Garland Publishing.

Burt, Jonathan (2002) *Animals in Film*, London: Reaktion Books.

Burton, Antoinette (1994) *Burdens of History: British Feminists, Indian Women, and Imperial Culture, 1865-1915*, Chapel Hill and London: University of North Carolina Press.

Busch, Briton (1985) *War Against the Seals: A History of the North-American Seal Fishery*, Kingston and Montreal: McGill-Queen's University Press.

Cabaton, A. (1911) *Java, Sumatra and the Other Islands of the Dutch East Indies*, London: T. Fisher Unwin.

Callwell, C. E. (1990) *Small Wars, a Tactical Textbook for Imperial Soldiers*, Novato: Presidio Press.

Campo, J. À (1992) *Koninklijke Paketvaart Maatschappij: Stoomvaart en Staatsvorming in de Indonesische Archipel, 1888-1914* [Royal Shipping Company, Steamshipping and State Formation in the Indonesian Archipelago], Hilversum: Verloren.

Cape of Good Hope Almanac and Annual Register (1845), Cape Town: A. S. Robertson.

Carlile, Nancy (1993) 'The Chewed Chair Leg and the Empty Collar: Mementos of Pet Ownership in New England', *Dublin Seminar for New England Folklife Annual Proceedings*, vol. 18 (pp. 130-146).

'Carruajes "La Union" ' [Carriages "La Union"] (1889) Philippine National Archives, Carruajes, Carros y Caballos, Bundle 7.

Cartmill, Matt (1993) *A View to a Death in the Morning: Hunting and Nature through History*, Cambridge: Harvard University Press.

Casey, H. L. (1906) Director of Studs to the Chief of the Bureau of Agriculture, Baguio, 31 August 1905. In *Report of the Philippine Commission, 1905*, Washington: Government Printing Office.

Cassidy, Rebecca (2002) *The Sport of Kings: Kinship, Class and Thoroughbred Breeding in Newmarket*, Cambridge: Cambridge University Press.

Catalan, Nemesio (1935) 'The Animal Problem of the Future Philippine Army', *Philippine Journal of Animal Industry*, vol. 2, no. 1 (pp. 67-77).

Census of the Philippine Islands, 1903 (1905) Washington: Bureau of the Census.
——, *1918* (1920-1921) Manila: Bureau of Printing.

Chambert-Loir, Henri (1994) 'State, City, Commerce: The Case of Bima', *Indonesia* 57 (pp. 71-88).

Charney, Michael (2004) *Southeast Asian Warfare, 1300-1900*, Leiden: Brill.

Chevenix-Trench, Charles (1980) *A History of Horsemanship*, London: Doubleday.

Child, Daphne (1967) *Saga of the South African Horse*, Cape Town: Howard Timmins.

Chirino, Pedro (1973) 'Relación de las Islas Filipinas, 1604'. In Blair and Robertson, *The Philippine Islands, 1493-1898*, vol. 12 (pp. 169-322).

Chivers, Keith (1983) 'The Supply of Horses in Great Britain in the Nineteenth Century'. In F. Thompson (ed.), *Horses in European Economic History: A Preliminary Canter*, Reading: The British Agricultural History Society (pp. 31-49).

Choisy, François-Timoléon, (1993) *Journal of a Voyage to Siam 1685-1686*, Oxford University Press.

Christie, Jan Wisseman (2004) 'The Agricultural Economies of Early Java and Bali'. In Boomgaard and Henley (eds), *Smallholders and Stockbreeders: Histories of Foodcrop and Livestock Farming in Southeast Asia*, Leiden: KITLV Press (pp. 47-68).

Clarence-Smith, William Gervase (1979) *Slaves, Peasants and Capitalists in Southern Angola, 1840-1926*, Cambridge: Cambridge University Press.

—— (2002) 'Horse Trading: The Economic Role of Arabs in the Lesser Sunda Islands, c. 1800 to c. 1940'. In Huub de Jonge and Nico Kaptein (eds), *Transcending Borders: Arabs, Politics, Trade and Islam in Southeast Asia*, Leiden: KITLV Press (pp. 143-162).

—— (2004a) 'Horse Breeding in Mainland Southeast Asia and Its Borderlands'. In Peter Boomgaard and David Henley (eds), *Smallholders and Stockbreeders: Histories of Foodcrop and Livestock Farming in Southeast Asia*, Leiden: KITLV Press (pp. 189-210).

—— (2004b) 'Elephants, Horses, and the Coming of Islam to Northern Sumatra', *Indonesia and the Malay World*, 2004, vol. 32 (pp. 271-284).

—— (2004c) 'Cape to Siberia: The Indian Ocean and China Sea Trade in Equids'. In David Killingray, Margaret Lincoln and Nigel Rigby (eds), *Maritime Empires: British Imperial Maritime Trade in the Nineteenth Century*, Woodbridge: Boydell and Brewer (pp. 48-67).

Clotet, José Maria (1973) Fr José Maria Clotet to Rev. Fr Rector of Ateneo Municipal, Talisayan, 11 May, 1889. In Blair and Robertson, *The Philippine Islands, 1493-1898*, vol. 43 (pp. 288-306).

Clutton-Brock, Juliet (1992) *Horse Power*, Cambridge: Harvard University Press.

Coates, Austin (1994) *China Races*, Hong Kong: Oxford University Press.

Commander of the GCV to the Director-General of Civil Administration, 4 September (1888) Philippine National Archives, Veterinarios.

Conservation of Early Domesticated Animals of Southern Africa' (1994) Conference proceedings, Willem Prinsloo Agricultural Museum, 3-4 March.

Coolhaas, W. Ph. (ed.) (1964, 1968, 1971, 1975, 1976, 1979, 1985) *Generale Missiven van Gouverneurs-Generaal en Raden aan Heren XVII der Verenigde Oostindische Compagnie* [General Letters from the Governors-General and Council to the Gentlemen Seventeen of the United East India Company], Vols. II, III, IV,V, VI, VII, VIII, The Hague: Martinus Nijhoff.

Corchera, Sebastian Hurtado de (1973) Letter to Felipe IV, 30 June, 1636. In Blair and Robertson (eds), *The Philippine Islands, 1493-1898*, vol. 26 (pp. 60-264).

Cortesão, Armando (ed.) (1944) *The Summa Oriental of Tomé Pires: An Account of the East, from the Red Sea to Japan, Written in Malacca and India in 1512-1515,* London: Hakluyt Society.

—— (ed.) (1990) *The Suma Oriental of Tomé Pires*, New Delhi & Madras: Asian Educational Services.

Couperus, G.W. (1886) 'Les Races Chevalines des Iles de la Sonde' [Breeds of Horses of the Sunda Islands], *Revue Coloniale Internationale* 2 (pp. 29-38).

Courtellemont, Gervais (1904) *Voyage au Yunnan* [Voyage to Yunnan], Paris: Plon-Nourrit.

Cowper Rose (1829) *Four Years in Southern Africa*, London: Colburn and Bentley.

Crawfurd, John (1856) *A Descriptive Dictionary of the Indian Islands & Adjacent Countries*, London: Bradbury & Evans.

—— (1971) *A Descriptive Dictionary of the Indian Islands & Adjacent Countries*, Kuala Lumpur: Oxford University Press.

Cribb, Robert (1993) 'Development Policy in the Early 20th Century'. In Dirkse, J.P., Frans Hüsken and Mario Rutten, *Development and Social Welfare. Indonesia´s Experiences under the New Order*, Leiden: KITLV Press [Verhandelingen 156], pp. 224-245.

Crisp, Olga (1983) 'Horses and Management of a Large Agricultural Estate in Russia at the End of the Nineteenth Century'. In Francis Thompson (ed.), *Horses in European Economic History: A* Preliminary Canter, Reading: The British Agricultural History Society (pp. 156-176).

Crist, Eileen (1999) *Images of Animals: Anthropomorphism and Animal Mind*, Philadelphia: Temple University Press.

Cronon, William (1992) 'A Place for Stories: Nature, History, and Narrative', *Journal of American History*, vol. 78, no. 2 (pp. 1347-1376).

Croo, Maurice Henri du (1917) 'De Verbetering van den Paardenstapel op het Eiland Soembawa' [The Improvement of the Horse Population on the Island of Sumbawa], *Koloniale Studiën*, vol. 1 (part II) (pp. 479-491).

Crosby, Alfred (1972) *The Columbian Exchange: Biological Consequences of 1492*, Westport: Greenwoods.

Crosby, Alfred W. (1986) *Ecological Imperialism: The Biological Expansion of Europe, 900 to 1900*, Cambridge, Cambridge: Cambridge University Press.

Cuneo, Pia (2000) 'Beauty and the Beast: Art and Science in Early Modern European Equine Imagery', *Journal of Early Modern History*, vol. 4, nos. 3-4 (pp. 269-321).

Curtin, Philip (1989) *Death by Migration: Europe's Encounters with the Tropical World in the Nineteenth Century*, Cambridge: Cambridge University Press.

Daghregister Gehouden int Casteel Batavia vant Passerende daer ter Plaetse als over Geheel Nederlandts-India [Daily Register Kept in Castle Batavia of Occurences There and in the Entire Netherlands Indies] (1887-1931), 31 vols (covering selected years between 1624 and 1682), Batavia/'s-Gravenhage: Landsdrukkerij/Nijhof.

Dalby, Simon (2004) 'Ecological Politics, Violence, and the Theme of Empire', *Global Environmental Politics*, vol. 4, no. 2, May (pp. 1-11).

Dam, Pieter van (1927-1954) *Beschrijvinge van de Oostindische Compagnie* [Descriptions of the East Indian Company], Gravenhage: Martinus Nijhoff: Rijks Geschiedkundige Publicatiën, edited by F.W. Stapel (first 6 vols) and C. W. Th. van Boetzelaer.

Dames, Mansel Longworth (ed.) (1918) *The book of Duarte Barbosa: An Account of the Countries Bordering on the Indian Ocean and their Inhabitants,*

Written by Duarte Barbosa and Completed about the Year 1518 A.D., London: Hakluyt Society.

Dampier, William (1931) *Voyages and Discoveries* [1699], edited by C. Wilkinson, London: The Argonaut Press.

—— (1939) *A Voyage to New Holland*, edited by J. A. Williamson, London: Argonaut Press.

—— (1973) 'A New Voyage Round the World, 1703'. In Blair and Robertson, *The Philippine Islands, 1493-1898*, vol. 38 (pp. 241-285) and vol. 39 (pp. 21-121).

Darwin, Charles (1872) *The Expression of the Emotions of Man and Animals*, London: Murray.

Dasmariñas, Gómez Pérez (1973) to the King of Camboja, 27 September, 1593. In Blair and Robertson (eds), *The Philippine Islands, 1493-1898*, vol. 9 (pp. 76-78).

Davenport Adams, W.H. (1880) *The Eastern archipelago*, London: T. Nelson and Sons.

Dawkins, Marian (1993) *Through Our Eyes Only? The Search for Animal Consciousness*, Oxford: W. H. Freeman Spektrum.

De Residentie Kadoe naar de Uitkomsten der Statistieke Opname [The Residency of Kadoe Based on the Results of the Statistical Survey] (1871), Batavia: Landsdrukkerij.

Delort, Robert (1984) *Les Animaux ont une Histoire*, Paris: Éditions du Seuil.

Derry, Margaret (2003) *Bred for Perfection: Shorthorn Cattle, Collies, and Arabian Horses since 1800*, Baltimore: Johns Hopkins University Press.

Desquibel, Pedro (1973) 'Reception of the Royal Seal at Manila 8 June, 1598'. In Blair and Robertson (eds), *The Philippine Islands, 1493-1898*, vol. 10 (pp. 133-139).

Dhiravat na Pomberja (2001) *Siamese Court Life in the Seventeenth Century as Depicted in European Sources*, Bangkok: Chulalongkorn University.

Diamond, Jared (1997) *Guns, Germs, and Steel: The Fates of Human Societies*, New York: W.W. Norton.

Diaz, Carlos (1936) 'Breeding Performance of Arabian Stallions of the Bureau of Animal Industry', *Philippine Journal of Animal Industry*, vol. 3, no. 4 (pp. 263-272).

Dijk, G. C. A. (1904) 'De Sandelhout en de Paardenfokkerij in Nederlandsch Indië', *De Indische Gids*, vol. 26, no. 1, ii (pp. 367-94).

Dobbin, Christine (1983) *Islamic Revivalism in a Changing Peasant Economy: Central Sumatra, 1784-1847*, London, Malmö: Curzon Press.

Doniger, Wendy (1999) 'Presidential Address: "I Have Scinde": Flogging a Dead (White Male Orientalist) Horse', *Journal of Asian Studies*, vol. 58, no. 4 (pp. 940-960).

Dove, Michael (1984) 'Man, Land, and Game in Sumbawa, Eastern Indonesia', *Singapore Journal of Tropical Agriculture*, vol. 5, no. 2 (pp. 112-124).

Drewer, M. S. (1969) 'The Domestication of the Horse'. In Peter Ucko and George Dimbleby (eds), *The Domestication and Exploitation of Plants and Animals*, London: Duckworth (pp. 471–478).

Du Plessis, S. F. (1934) *Die Verhaal van die Dagverhaal van Jan van Riebeeck* [The Narrative of the Diary of Jan van Riebeeck], D. Litt. thesis: University of Pretoria.

Ducray, Charles G. et al. (1938) *Ile Maurice*, Port Louis: General Printing and Stationery Co. Ltd.

Dumarçay, Jacques (1991) *The Palaces of South-East Asia*, Singapore: Oxford University Press.

Dunlap, Thomas (1999) *Nature and the English Diaspora – Environment and History in the United States, Canada, Australia and New Zealand*, Cambridge: Cambridge University Press.

'Early Franciscan Missions [1649]' (1973). In Blair and Robertson (eds), *The Philippine Islands, 1493-1898*, vol. 35 (pp. 278-322).

Edwards, C. (1916) 'The Live-stock Industry of the Philippines', *Philippine Agricultural Review*, vol. 9, no. 2 (pp. 136-149).

Edwards, Peter (1988) *The Horse Trade of Tudor and Stuart England*, Cambridge: Cambridge University Press.

El Comandante Jefe to Corregidor de Manila (1884), Philippine National Archives, Animales Sueltos, Corregimiento de Manila.

Elder, G. J. Wolch, and J. Emel. (1998) 'Race, Place, and the Bounds of Humanity', *Society & Animals*, vol. 6, no. 2 (pp. 183-202).

Eldridge, E. (1993) *A South African Kingdom – The Pursuit of Security in Nineteenth Century Lesotho*, Cambridge: Cambridge University Press.

Ellenberger, D. F. (1969) *History of the Basuto*, New York: Negro University Press.

Elliott, Charles B. (1968) *The Philippines to the End of the Commission Government*, New York: Greenwood Press.

Elphick, Richard (1977) *Kraal and Castle: Khoikhoi and the Founding of White South Africa*, New Haven: Yale University Press. (pp. 90-110).

Emel, Jody (1998) 'Are You Man Enough, Big and Bad Enough? Wolf Eradication in the US'. In Wolch and Emel, *Animal Geographies* (pp. 91-116).

Encyclopaedie van Nederlandsch-Indië, (1917-1921) The Hague: W. P. van Stockum & Son.

Epstein, H. (1971) *The Origin of the Domestic Animals of Africa*, New York: Africana.

'Espediente Original, Pampanga 1859' [The First or Original Proceedings] (1851) Philippine National Archives, Carreras de Caballos, Bundle 1.

Estadística Demografico-Sanitaria del Radio Municipal de L. M. I. Y. S. L.C. de Manila de Enero a Diciembre de 1886, Defunciones [Demographic and Health Statistics of the Municipal Authority of Manila from January to December 1886, Deaths] (1888), *El Comercio*, 3 March.

'Estado del Numero de Carruages, Carros y Caballos, Varias Provincias: Bohol, Camarines Norte, Capiz, Cebu, Iloilo, Intramuros, La Union, Manila, Negros, Tarlac, Zambales' [The Number of Carriages, Carts and Horses, Various Provinces] (1889) Philippine National Archives, Carruajes, Carros y Caballos, Bundle 7.

'Expediente sobre Cumplimiento de la Real Orden No 275 De 30 Marzo de 1885 para el Fomento y Meyora de la Raza Caballar en las Islas, 1 May' [Proceedings on the Implementation of the Royal Order No.275 of 30 March 1885 to Promote and Improve the Breed of Horse in the Islands] (1888), Philippine National Archives, Raza de Caballeria de Filipinas, Formento y Mejora.

'Expediente sobre Dejar Sin Efecto la Circular de Este Centro, de 27 de Julio de 1886' [Proceedings on Leaving in Effect the Circular of this Centre, of 27 July 1886] (1887) Philippine National Archives, Carruajes, Carros y Caballos, Bundle 7.

'Expediente sobre la Ejecución del Proyecto de Renovación y Mejora de la Cria Caballar de este Archipiélago' [Proceedings on the Management of the Project to Reinvigorate and Improve the Breed of Horse of this Archipelago] (1881) Philippine National Archives, Raza de Caballeria, Formento y Mejora.

'Expediente sobre Reclamación, Iloilo 1884' [Proceedings about the Complaint, Iloilo 1884] (1884) Philippine National Archives, Carruajes, Carros y Caballos, Bundle 7.

Fattah, Hala M. (1997) *The Politics of Regional Trade in Iraq, Arabia and the Gulf, 1745-1900*, Albany (New York): State University of New York Press.

Felipe II (1973) Instructions from Felipe II to Governor Gómez Pérez Dasmariñas, 9 August, 1589. In Emma Blair and Alexander Robertson, *The Philippine Islands, 1493-1898*, Mandaluyong: Cachos Hermanos, vol. 7 (pp. 141-172).

—— (1973) The Audiencia of Manila Re-established, 26 November 1595. In Blair and Robertson (eds), *The Philippine Islands, 1493-1898*, vol. 9 (pp. 189-191).

Felipe Jovantes to Guardia Civil Veterana (1874), Philippine National Archives, Veterinarios.

Ferrars, Max and Bertha (1900) *Burma*, London: Sampson Low, Marston & Co.

Fielding, Denis, and Krause, Patrick (1998) *Donkeys*, London: Macmillan Education.

Firth, Raymond (1951) *Elements of Social Organization*, London: Watts.

Flannery, Tim (2001) *The Eternal Frontier: An Ecological History of North America and Its Peoples*, London: William Heinemann.

'Floods in the Philippines 1691-1911', *Archives of the Manila Observatory*, Box 10, 37.

Foreman, John (1899, 1985) *The Philippine Islands*, Mandaluyong: Cacho Hermanos.

—— (1890) *The Philippine Islands,* London: Sampson Low, Marston, Searle & Rivington.

—— (1906) *The Philippine Islands*, Shanghai: Kelly and Walsh.

'Formación Previa en Averiguación' [Review of the Case] (1874) Philippine National Archives, Veterinarios.

Foronda, Marcelino (1986) *Insigne y Siempre: Leal Essays on Spanish Manila*, Manila: De La Salle University, History Department and the Research Center.

Foster, William (ed.) (1943) *The Voyage of Sir Henry Middleton to the Moluccas 1604-1606*, London: Hakluyt Society (Works issued by the Hakluyt Society, 2nd series, 88).

Fox, James (1977) *Harvest of the Palm: Ecological Changes in Eastern Indonesia*, Cambridge, London: Harvard University Press.

Francis, Emanuel (1856) *Herinneringen uit de Levensloop van een Indisch Ambtenaar van 1815 tot 1851* [Memories from the Course of the Life of an 'Indisch' Official from 1815 to 1851], Batavia: Van Dorp. 3 vols.

Frankel, P. H. (1984) *Pretoria's Praetorians, Civil-military Relations in South Africa*, Cambridge: Cambridge University Press.

Franklin, Adrian (1999) *Animals and Modern Cultures: A Sociology of Human-Animal Relations in Modernity*, London: Sage.

Freijss, J. P. (1859) 'Schets van den handel van Sumbawa' [Sketch of the Trade on Sumbawa], *Tijdschrift van Nederlandsch-Indië* (TNI) II (pp. 268-283).

French, Richard (1975) *Antivivisection and Medical Science in Victorian Society*, Princeton: Princeton University Press.

Fudge, Erica (2000) *Perceiving Animals: Humans and Beasts in Early Modern English Culture*, Basingstoke: Macmillan Press.

Geoffroy Saint-Hilaire, Isidore (1861) *Acclimatation et Domestication des Animaux Utiles* [Acclimatization and Domestication of Work Animals], Paris.

Generale Missiven van Gouverneurs-Generaal en Raden aan Heren XVII der Verenigde Oostindische Companie [General Missives of the Governors-General and Council to the Gentlemen XVII of the United East Indian Company] (1960-2004), edited by W. P. Coolhaas et al, 's-Gravenhage: Nijhoff; Rijks Geschiedkundige Publicatiën, 11 vols.

Gervaise, Nicolas (1989) *The Natural and Political History of the Kingdom of Siam*, trans. John Villiers, Bangkok: White Lotus.

Ginsburg, Henry (1989) *Thai Manuscript Painting*, London: The British Library.

Glick, Thomas (1974) 'Spain'. In Thomas Glick (ed.), *The Comparative Reception of Darwin*, Austin and London: University of Texas Press (pp. 307-345).

Glukin, Neil Dana (1977) 'Pet Birds and Cages of the 18th Century', *Early American Life*, vol. 8 (pp. 38-59).

Goethals, Peter R. (1961) *Aspects of Local Government in a Sumbawan Village (Eastern Indonesia)*, Ithaca, New York: Cornell University. [Monograph Series. Modern Indonesia Project, Southeast Asia Program, Department of Far Eastern Studies].

Gommans, Jos (1994) 'The Horse Trade in Eighteenth-century South Asia', *Journal of the Economic and Social History of the Orient*, vol. 37 (pp. 228-50).

González y Martín, R. (1986) *Filipinas y Sus Habitantes*, Béjar: Establecimiento Tipográfico de la Viuda de Aguilar.

Goodall, Daphne Machin (1968) *Paarderassen van de Wereld* [Horses of the World], Zwolle: La Rivière en Voorhoeve.

Goody, Jack (2000) *The Power of Written Tradition*, Washington: Smithsonian Institute Press.

Goor, Jurriën van (ed.) (1988) *Generale Missiven van Gouverneurs-Generaal en Raden aan Heren XVII der Verenigde Oostindische Compagnie*, Vol. IX, The Hague: Martinus Nijhoff.

Gordon, Lady Duff (1925) *Letters from the Cape*, Cape Town: Maskew Miller.

Great Britain, Foreign Office, Historical Section (1919) *Dutch Timor and the Lesser Sunda Islands*, London.

—— (1920) *Portuguese Timor*, London.

Green, George (1947) 'How to Train a Horse to Triple', *Farmers Weekly*, vol. 74, no. 55, 24 December 1947.

Grimbeek, Christiaan (1985) '*Die Totstandkoming van die Unieverdegingsmag met Spesifieke Verwysing na die Verdedigingswette van 1912 and 1922*'[The Establishment of the Union Defence Force with Specific Reference to the Defence Acts of 1912 and 1922] D.Phil thesis: University of Pretoria.

Groeneveld, W. (1916) 'Het Paard in Nederlandsch-Indië; Hoe het is Ontstaan, hoe het is en Hoe het kan Worden' [The Horse in the Netherlands Indies; How it Originated, How it is, and How it could Become], *Veeartsenijkundige Bladen voor Nederlandsch-Indië*, vol. 28 (pp. 197-237).

Groeneveld, W. and J. C. Witjens (1924) *Het paard* [The Horse], Haarlem: Tjeenk Willink. [Onze Koloniale Dierenteelt. Populaire handboekjes over Nederlandsch-Indische nuttige dieren].

Groeneveldt, Willem Pieter (1880) 'Notes on the Malay Archipelago and Malacca, compiled from Chinese Sources', *Verhandelingen Bataviaasch Genootschap*, 1880, vol. 39 (pp. 34-39).

Groves, Colin P. (1974) *Horses, Asses and Zebras in the Wild*, Sanibel: Ralph Curtis Books.

Guillemard, F.H.H. (1889) The Cruise of Marchesa to Kamschatka & New Guinea (with Notices of Formosa, Liu-Kiu, and Various Islands of the Malay Archipelago), London: John Murray. [Second edition].

Haan, Frederik de (1910-12) *Priangan: De Preanger-Regentsschappen onder het Nederlandsche Bestuur tot 1811* [Priangan: The Preanger Regencies under Dutch Administration up to 1811], Batavia: Kolff, 4 vols.

Haggard, H. Rider *Jess* (1887, 1909), London: Smith, Elder.

—— *Finished* (1917, 1962), London: Macdonald.

Hamilton, Jill (2000) *Marengo: The Myth of Napoleon's Horse*, London: Fourth Estate.

Haraway, Donna (1989) *Primate Visions: Gender, Race, and Nature in the World of Modern Science*, New York and London: Routledge.

Harris, Marvin (1965) 'The Myth of the Sacred Cow'. In Anthony Leeds and
    Andrew P. Vayda (eds) *Man, Culture and Animals,* Washington: American
    Association for the Advancement of Science (pp. 217-228).
Harrison, Mark (1996) '"The Tender Frame of Man": Disease, Climate, and Racial
    Difference in India and the West Indies, 1760-1860', *Bulletin of the History
    of Medicine,* vol. 70 (pp. 68-93).
Harvey, Godfrey E. (1967) *History of Burma,* London: Frank Cass.
Hasselt, Arend Ludolf van (1882) *Volksbeschrijving van Midden-Sumatra* [Descrip-
    tion of the People of Central Sumatra] (vol. 3 of P. J. Veth (ed.) *Midden-Sumatra.
    Reizen en Onderzoeking der Sumatra-Expeditie 1877-1879* [Central Sumatra.
    Travels and Research of the Sumatra Expedition 1877-1879], Leiden: Brill.
Hayes, M. H. (1900) *Among Horses in South Africa,* London: Everett.
Henderikus, J. Stam and Tanya Kalmanovitch (1998) 'E. L. Thorndike and the
    Origins of Animal Psychology; On the Nature of the Animal in Psychology',
    American Psychologist, vol. 53, no. 10 (pp. 1138-1142).
Henninger-Voss, Mary (2002) *Animals in Human Histories: The Mirror of Nature
    and Culture,* Rochester: University of Rochester Press, 2002.
Heshusius, C. A. (c. 1978) *KNIL-Cavalerie, 1814-1950: Geschiedenis van de
    Cavalerie en Pantsertroepen van het Koninklijk Nederlands-Indische Leger*
    [History of the Light and Heavy Cavalry of the Royal Netherlands-Indies
    Army], [no place].
Hitchcock, Michael J. (1983) *Technology and Society in Bima, Sumbawa, with special
    reference to House Building and Textile Manufacture,* Oxford: University of
    Oxford. [PhD thesis, Department of Ethnology and Prehistory].
Hoekstra, Pieter (1948) 'Paardenteelt op het Eiland Soemba' [Horsebreeding on the
    Island of Sumba], Batavia [PhD thesis, Universiteit van Indonesië, Batavia].
—— (1948) *Paardenteelt op het Eiland Soemba* [Horsebreeding on the Island of
    Sumba], Batavia: Drukkerij V/H John Kappee.
Hoen, H. 't (1919) *Veerassen en Veeteelt in Nederlandsch-Indië* [Livestock Breeds
    and Stockbreeding in the Netherlands Indies] Weltevreden: Kolff.
Hogendorp, Willem van (1779-1780) 'Beschryving van het Eiland Timor, voor
    Zoo Verre het tot Nog toe Bekend is' [Description of the Island of Timor,
    in So Far as it is Known till Now], *Verhandeling Bataviaasch Genootschap,*
    vol. 2.
Ileto, Reynaldo (1998) 'Hunger in Southern Tagalog, 1897-1898'. In Reynaldo
    Ileto, *Filipinos and their Revolution: Event, Discourse, and Historiography,*
    Quezon City: Ateneo de Manila Press.
Ingold, Tim (1980) *Hunters, Pastoralists and Rancher Reindeer Economies and
    Their Transformations,* Cambridge: Cambridge University Press.
Instituut voor Nederlandse Geschiedenis, *VOC-Glossarium* [VOC glossary] (2000)
    Den Haag: Instituut voor Nederlandsche Geschiedenis.
'Inventario de los Cinco Caballos Arabes del Estado que Existen en este Gobierno
    que por Orden del Exmo Sor. Gobor. Gral. del Archipielago se Remiten a

Disposicion del Ecmo. Sor. Director Gral. de Admon. Civil por Conducto del Vapor 'Elcano', 12 February' [Record of the Five Arab Stallions Currently in the Possession of the State that by Order of the Governor-general of this Archipelago are placed at the Disposition of the Director-general of Civil Administration by Means of the Steamship 'Elcano'] (1890), Philippine National Archives, Raza de Caballeria de Filipinas, Formento y Mejora.

Isenberg, Andrew (2000) *The Destruction of the Bison*, Cambridge: Cambridge University Press.

Ito, Takeshi (1984) 'The World of the Adat Aceh, a Historical Study of the Sultanate of Aceh', PhD thesis, Australian National University.

[Jaarboek] (1907-1930) *Jaarboek van het Departement van Landbouw, Nijverheid en Handel in Nederlandsch-Indië* [Year-book of the Department of Agriculture in the Netherlands-Indies] [24 volumes, covering selected years between 1906 and 1929], Batavia: Landsdrukkerij; Weltevreden: Albrecht.

[Jaarverslag] (1923-1941) *Jaarverslag van de Burgerlijke Veeartsenijkundige Dienst (BVD)* [Yearly report of the Civil Veterinary Service] [29 volumes, covering selected years between 1922 and 1940], Batavia: Departement van Economische Zaken.

Jacottet, E. (1908) *The Treasury of Ba-Suto Lore*, Morija: Sesuto Book Depot.

Jardine, Lisa and Jerry Brotton. (2000) *Global Interests: Renaissance Art between East and West*, Ithaca: Cornell University Press.

Jasper, J.E. (1908) 'Het Eiland Soembawa en zijn Bevolking' [The Island of Sumbawa and its Population], *Tijdschrift voor het Binnenlandsch Bestuur*, vol. 34 (pp. 60-147).

Jennings, Eric (1970) *The Singapore Turf Club*, Singapore: Singapore Turf Club.

Jones, A. M. Barrett (1984) *Early Tenth Century Java from Inscriptions*, Dordrecht, Cinnaminson: Foris (Verhandelingen KITLV, 107).

Jones, Susan (2003) *Valuing Animals: Veterinarians and Their Patients in Modern America*, Baltimore: Johns Hopkins University Press.

Jong Boers, Bernice de (1994) 'Tambora 1815: De Geschiedenis van een Vulkaanuitbarsting in Indonesië' [Tambora 1815: The History of a Volcanic Eruption in Indonesia], *Tijdschrift voor Geschiedenis*, vol. 107, no. 3 (pp. 371-392).

—— (1995) 'Mount Tambora in 1815: A Volcanic Eruption in Indonesia and its Aftermath', *Indonesia*, vol. 60 (pp. 36-60).

—— (1995) 'Horses: One of the First Commodities of the Island of Sumbawa (Indonesia)', Yogyakarta/Leiden Gadjah Mada University/Leiden University. [Paper written for the first Summer Course in Indonesian Modern Economic History, unpublished].

—— (1997) 'Some Notes on the History of Livestock in Indonesia (with a Special Focus on Sumbawa)' [Paper written for the conference on 'Animals in Asia: relationships and representations', 15-16 September 1997, unpublished].

—— (1997) 'Paardenfokkerij op Sumbawa (1500-1930)' [Horse Breeding on the Island of Sumbawa (1500-1930)], *Spiegel Historiael*, vol. 32, no. 10/11 (pp. 438-443).

—— (2000/2001) 'Een Proefschrift in Wording over de Milieugeschiedenis van het Eiland Sumbawa, Indonesië' [A Note on Writing a PhD Thesis about the Environmental History of the Island of Sumbawa, Indonesia], *De Boekerij: Mededelingenblad van de Vereniging Vrienden der Bibliotheek van de Koninklijke Nederlandse Akademie van Wetenschappen*, vol. 5, no. 3/vol. 6, no. 1, (pp. 10-15).

Jonge, Jan Karel Jacob de, (& M. L. van Deventer) (1862-1895) *De Opkomst van het Nederlandsch Gezag in Oost-Indië. Versameling van Onuitgegeven Stukken uit het Oud-koloniaal Archief (1595-1814)* [The Rise of Dutch Power in the East Indies: Collection of Unpublished Documents from the Old-Colonial Archives (1595-1814)], 17 vols, 's-Gravenhage: Nijhoff.

José Sora to Sor. Corregidor de Manila (1878), Philippine National Archives, Animales Sueltos, Corregimiento de Manila.

Judd, B.C. (1938) 'With the C. M. R', *Nongquai*, June.

Junghuhn, Franz Wilhelm (1853-1854) *Java, zijne Gedaante, zijn Plantetooi en Inwendige Bouw* [Java, its Appearance, its Flora, and its Morphology], 3 vols, 's-Gravenhage: Mieling.

Kaempfer, Engelbert 91987) *A Description of the Kingdom of Siam 1690*, trans. J.G. Scheuchzer, Bangkok: White Orchid Press.

Kaur, Amarjit (1985) Bridge and Barrier, Transport and Communications in Colonial Malaya, 1870-1957, Singapore: Oxford University Press.

Kay, Robert (1939) 'Java Ponies and Others', The Horse (Illustrated), *The Quarterly Review of the Institute of the Horse and Pony Club*, vol. 11, no. 41 (pp. 24-7).

Kelly, Joan (1984) *Women, History and Theory: The Essays of Joan Kelly*, Chicago and London: University of Chicago Press.

Kemp, Pieter Hendrik, van der (1890) 'Historisch Overzicht van de Pogingen Aangewend tot Verbetering en Veredeling van het Paardenras in Nederlandsch-Indië' [Historical Overview of the Attempts to Improve the Breed of Horses in the Netherlands Indies], *Veeartsenijkundige Bladen voor Nerderlandsch-Indië,*1890, vol. 4 (pp. 327-387).

Kennedy, John (1992) *The New Anthropomorphism*, Cambridge: Cambridge University Press, 1992.

Kete, Kathleen (1994) *The Beast in the Boudoir: Pet-keeping in Nineteenth Century Paris*, Berkeley: University of California Press

Kidd, Jane (1985) *The Horse The Complete Guide to Horse Breeds and Breeding*, London: Longmeadow Press.

Kipling, J. L. (1921) *Beast and Man in India, a Popular Sketch of Indian Animals in their Relation with People*, London: Macmillan.

Kistermann, H. (1990) 'De Paardenhandel van Nusa Tenggara 1815-1941' [The Horse Trade in Nusa Tenggara], Amsterdam: Vrije Universiteit, [Unpublished doctoral thesis].

Klaits, Joseph and Barrie Klaits (eds) (1974) *Animals and Man in Historical Perspective*, New York: Harper & Row.

Knight, John (ed.) (2004) *Wildlife in Asia: Cultural Perspectives*, London and New York: Routledge Curzon.

Kniphorst, J. H. P. E. (1885) 'Een Terugblik op Timor en Onderhoorigheden' [A Review of Timor and its Dependencies], *Tijdschrift voor Nederlandsch Indië*, vol. 14, no. 1 (pp. 355-80, 401-83) and no. 2 (pp. 1-41, 81-146, 198-204, 241-311, 321-62).

Koert, E. (1912) 'Stock Breeding in the Catanduanes Islands', *Philippine Agricultural Review*, vol. 5 (pp. 305-308).

Kolbe, P. *The present state of the Cape of Good Hope, 1713*, trans. G. Medley, London: W. Innys and R. Manly.

Kolbe, Peter (1727) *Naaukeurige en Uitvoerige Beschryving van Kaap de Goede Hoop*, Amsterdam: B. Lakeman.

Koloniaal Verslag, (1849-) The Hague: M. Nijhoff.

Kretzer, David (1928) 'How to Build Up and Improve a Herd or Flock (With Description of their Most Common Diseases in the Philippines)', *Philippine Agricultural Review*, vol. 21 (pp. 215-330).

Krom Sinlapakon (Fine Arts Department) (1969) *Tamra phichai songkhram* [Manual on Warfare], Bangkok: Fine Arts Department.

—— (1978) Ruang kotmai tra sam duang [The Law of the Three Seals] Bangkok: Fine Arts Department.

—— (1987) Krabuan phayuhayatra sathonlamak samai somdet phra narai maharat chamlong chak ton chabab nangsu samud thai khong ho samud haeng chat [Royal Land Procession in King Narai's Reign from old Thai Manuscript in National Library], and Riu krabuan hae phayuhayatra thang chonlamak chak ton chabab nangsusamud thai khong ho samud haeng chat [Water Procession from King Narai's Reign] Bangkok: Fine Arts Department.

—— (1994) *Prachum phongsawadan phak thi 82. Ruang phra ratcha phongsawadan krung siam chak ton chabab khong british museum krung london* ['British Museum' Recension of the Ayutthaya Royal Chronicles] Bangkok: Fine Arts Department.

—— (2001) *Tamra ma khong kao kap tamra laksana ma* [Old Horse Manuals] Bangkok: Fine Arts Department.

Kuperus, Gerrit (1936) *Het Cultuurlandschap van West-Soembawa* [The Man-made Landscape of Western-Sumbawa], Groningen, Batavia: J.B. Wolters. [PhD thesis Rijksuniversiteit Utrecht].

—— (1938) 'Beschouwingen over de Ontwikkeling en den Huidigen Vormenrijkdom van het Cultuurlandschap in de Onderafdeeling Bima (Oost-Soembawa)' [Considerations on the Genesis of the Current Polymorphology in the Man-Made Landscape in the District Bima (Eastern Sumbawa)], *Tijdschrift van het Koninklijk Aardrijkskundig Genootschap (KNAG)*, vol. 55, no. 2 (pp. 207-239).

'La Aprehensión de 23 Individuos de la Provincia de la Laguna y Cavite, Manila, 15 June 1891' [The Detention of 23 Suspects from the Provinces of Laguna and Cavite, Manila, 15 June 1891] (1891), Philippine National Archives, Cuadrilleros, Bundle 1.

La Loubère, Simon de (1969) *The Kingdom of Siam*, Kuala Lumpur: Oxford University Press.

Lamarck, Jean Baptiste (1809) *Philosophie Zoologique*, 2 vols, Paris.

Landen, Robert G. (1967) *Oman since 1856*, Princeton: Princeton University Press.

Landry, Donna (2004) 'The Bloody Shouldered Arabian and Early Modern English Culture', *Criticism*, Winter (pp. 41-69).

Law, Robin (1980) *The Horse in West African History*, Oxford: Oxford University Press.

Lawrence, Elizabeth (1981) *Rodeo: An Anthropologist Looks at the Wild and the Tame*, Knoxville: University of Tennessee Press.

—— (1990) 'Rodeo Horses: the Wild and the Tame'. In R. Willis, *Signifying Animals - Human Meaning in the Natural World*, London: Routledge (pp. 222-235).

Lee, C. G. et. al (2000) 'A Review of the Jindo, Korean Native Dog', *Asian-Australasian Journal of Animal Sciences*, vol. 13, no. 3, 2000, (pp. 381-389).

Legarda, Benito (1999) *After the Galleons: Foreign Trade, Economic Change and Entrepreneurship in the Nineteenth-Century Philippines*, Quezon City: Ateneo de Manila.

Leibbrandt, H. C. V. (1900) *Precis of the Archives of the Cape of Good Hope: Letters Received, 1695-1708*, Cape Town: Richards, vol. I, 1898, vol. II.

Lévi-Strauss, Claude (1963) *Totemism*, Boston: Beacon Press.

Lewin, Roger (1994) 'I Buzz Therefore I Think', *New Scientist*, 15 January (pp. 29-32).

Lichtenstein, H. (1928-1930) *Travels in Southern Africa in the Years 1803, 1804, 1805 and 1806*, Cape Town: The Van Riebeeck Society.

Lichtenstein, W. H. C. (1973) *Foundation of the Cape, About the Becuanas: Being a History of the Discovery and Colonisation of South Africa* [1811, 1807], trans. O. H. Spohr, Cape Town: Balkema.

Ligtvoet, Albertus (1876) 'Aanteekeningen Betreffende den Economischen Toestand en de Ethnographie van het Rijk van Soembawa' [Notes Concerning the Economic Situation and the Ethnography of the Realm of Sumbawa], *Tijdschrift voor Indische Taal-, Land- en Volkenkunde (TBG)* vol. 23 (pp. 555-592).

Limakatso Kendall, K. (1995) *Basali! Stories by and about women in Lesotho*. Pietermaritzburg: University of Natal Press.

Linn, Brian McAllister (2000) *The Philippine War 1899-1902*, Lawrence, Kansas: University of Kansas Press.

'Live Stock and Poultry in the Philippines; Horses' (1911) *The Philippine Agricultural Review*, vol. 4 (pp. 476-83).

Livingstone, David (1857) *Missionary Travels and Researches in South Africa*, London: Murray.

Livingstone, David (1987) 'Human Acclimatization: Perspectives on a Contested Field of Inquiry in Science, Medicine and Geography', *History of Science*, vol. 25 (pp. 359-394).

—— (1999) 'Tropical Climate and Moral Hygiene; The Anatomy of a Victorian Debate', *The British Journal for the History of Science*, vol. 32 (pp. 93-110).

Lombard, Denys (1967) *Le Sultanat d'Atjéh au Temp d'Iskandar Muda 1607-1636* [The Sultanate of Aceh in the Times of Iskandar Muda 1607-1636], Paris: Ecole Française d'Extrême-Orient.

—— (ed.) (1996) *Mémoires d'un Voyage aux Indes Orientales 1619-1622* [Memoires of a Voyage to the East Indies, 1619-1622]: *Augustin de Beaulieu, un Marchand Normand à Sumatra* [no pl.], Paris: Maisonneuve & Larose.

Loomba, Ania (1998) *Colonialism/Postcolonialism*, London and New York: Routledge.

Lush, J. L. *The Genetics of Populations*, Mimeo: University of Iowa Press, 1948; see 'From Jay L. Lush to Genomics: Visions for Animal Breeding and Genetics', Conference Proceedings, May 16-18, 1999, Department of Animal Science, Iowa State University.

Lye, William (ed.) (1975) *Andrew Smith's Journal of his Expedition into the Interior of South Africa, 1834-36*, Cape Town: A. A. Balkema, October 1834.

Macknight, C. C. (1980) 'Outback to Outback: The Indonesian Archipelago and Northern Australia'. In J. J. Fox (ed.), *Indonesia, the Making of a Culture*, Canberra: Australian National University (pp. 137-48).

Magner, Lois (1992) *A History of Medicine*, New York: Marcel Dekker Inc.

Mahamakut Ratchawitthayalai [Mahamakut Buddhist University] (1982) *Phra Traibidok Lae Atthakatha Thai* [The Tripitaka], Bangkok: Mahamakut Buddhist University.

Malcolmson, Robert and Stephanos Mastoris (1998) *The English Pig: A History*, London and Rio Grande: Hambledon Press.

Mallat, Jean (1983) *The Philippines: History, Geography, Customs, Agriculture, Industry and Commerce of the Spanish Colonies in Oceania*, Manila: National Historical Institute.

Manila Jockey Club, The (1997) *The Manila Jockey Club: 130 Years of Horse Racing in Southeast Asia*, Manila: Manila Jockey Club.

Manning, Aubrey and James Serpell (1994) *Animals and Human Society: Changing Perspectives*, London: Routledge.

Manuel, Mauro, Mario Tongson, Teodulo Topacio and Grace de Ocampo (eds.) (2002), *A Century of Veterinary Medicine in the Philippines*, Quezon City: University of the Philippines Press.

Manuel de Vos to Gobernador Civil de Provincia de Manila (1896) Philippine National Archives, Animales Sueltos, Corregimiento de Manila.

Marche, Alfred (1905, 1968) *Luzon and Palawan*, Manila: Filipiniana Book Guild.

Marks, Shula (1972) 'Khoisan Resistance to the Dutch in the Seventeenth and Eighteenth Centuries', *Journal of African History.* 13 (pp. 55-80).

Marsden, William (1975) *The History of Sumatra*, Kuala Lumpur, etc.: Oxford University Press.

Martin, Minnie (1903) *Basutoland - Its Legends and Customs*, London: Nichols.

Mas, Sinibaldo de (1843) *Informe sobre el Estado de las Islas Filipinas En 1842* [Report on the State of the Philippine Islands in 1842], Madrid: Imprenta de L. Sancha.

Mason, I. L., and Maule, J. P. (1960) *The Indigenous Livestock of Eastern and Southern Africa*, Farnham: Commonwealth Agricultural Bureaux.

Mason, Jeffrey and Susan McCarthy (1995) *When Elephants Weep: The Emotional Lives of Animals*, New York: Delta.

Mauritius Almanac and Colonial Register, 1926-27, Port Louis: General Printing and Stationery Co.

Maximo Loilla to Director-General de Administración Civil (1889) Philippine National Archives, Carruajes, Carros y Caballos, Bundle 7.

McClintock, Anne (1995) *Imperial Leather: Race, Gender and Sexuality in the Colonial Context*, New York and London: Routledge.

McEvoy, Arthur (1986) *The Fisherman's Problem: Ecology and the Law in the California Fisheries 1850-1980*, Cambridge: Cambridge University Press.

McLennan, Marshall (1980) *The Central Luzon Plain: Land and Society on the Inland Frontier*, Quezon City: Alemar-Phoenix Publishing House.

McNeill, William and John R. McNeill (2003) *The Human Web: A Bird's-Eye View of World History*, New York: W.W. Norton.

Melville, Elinor (1994) *A Plague of Sheep: Environmental Consequences of the Conquest of Mexico*, Cambridge: Cambridge University Press.

'Memoria de Ramon' (1893) Philippine National Archives, Servicios de Agricultoras, Cebu 1890-97.

'Memoria Preliminar para la Ejecución del Proyecto de Renovación y Mejora de la Cria Caballar de Este Archipiélago' [Preliminary Report on the Management of the Project to Reinvigorate and Improve the Breed of Horse of this Archipelago] (1883) Philippine National Archives, Raza de Caballeria, Formento y Mejora.

'Memoria sobre un Tranvia, Cebu' (1893) Philippine National Archives, Servicios de Agricultoras, Cebu 1890-97.

Memorieën van Overgave (MvO) [Memoranda of Transfer] The Nationaal Archief in The Haque, archives of the Ministerie van Kolonieën after 1900; MMK 340: MvO, C.H. van Rietschoten, Timor en Onderhorigheden [Timor and Dependencies], 25-7-1913; MMK 342: MvO, C. Schultz, Timor en Onderhorigheden [Timor and Dependencies], 13-6-1927; MMK 343: MvO, P.F.J. Karthaus, Timor en Onderhorigheden [Timor and Dependencies], 6-5-1931; MMK 344: MvO, E.H. de Nijs Bik, Timor en Onderhorigheden [Timor and Dependencies], 16-6-1934; MMK 34: MvO, J.J. Bosch, Timor en Onderhorigheden [Timor and Dependencies], 29-3-1938; KIT 1213, Militaire Memorie over de Afdeeling Soembawa [Military Report on the Region of Sumbawa], anonymous, 1929; KIT 1268, MvO, A. Couvreur, Timor en Onderhorigheden [Timor and Dependencies], 21-7-1924.

Mentzel, O. F. (1920) *Life at the Cape in the Mid-Eighteenth Century* [1784], trans. M. Greenless, reprinted Cape Town: Darter Brothers (pp. 146-156).

Merkens, Jan (1923) *De Veeteelt in Nederlandsch-Indië* [Cattle Breeding in the Netherlands-Indies] [Veeartsenijkundige Mededeeling, no. 47].

—— (1924) 'De Grootveestapel van Nederlandsch-Indië' [The Livestock Population of the Netherlands-Indies], *Koloniale Studiën*, vol. 8 (part I) (pp. 189-219).

Midgley, Mary (1984) *Animals and Why They Matter*, Athens: Georgia University Press.

Miller, Hugo H. (1920) *Economics Conditions in the Philippines*, Boston: Ginn & Co.

Mitchell, B. R. (1998) *International Historical Statistics*, London: Macmillan.

Mitchell, Robert, Nicholas Thompson and H. Lyn Miles (eds) (1997) *Anthropomorphism, Anecdotes and Animals*, Albany: State University of New York Press.

Mitzmain, M. (1912) 'Collected Notes on the Insect Transmission of Surra in Carabaos', *Philippine Agricultural Review*, vol. 5, no. 12 (p. 679).

Montero y Vidal, José (1886) *El Archipiélago Filipino y las Islas Marianas, Carolinas y Palaos Su Historia, Geografía y Estadística* [The Philippine Archipelago and the Marianas, Carolinas and Palau Islands Their History, Geography and Statistical Profile] Madrid: Imprenta y Fundición de Manuel Tello.

—— (1895) *Historia General de Filipinas, desde el Descubrimiento de Dichas Islas Hasta Nuestros Días* [A General History of the Philippines, Since the Discovery of the Said Islands until Nowadays], Madrid: Establecimiento Tipográfico de la Viuda é Hijos de Tello.

Moor, J. H. (1968 reprint of 1837 edn) *Notices of the Indian Archipelago and Adjacent Countries*, London: Oxford University Press.

Moree, P. J. (1998) *A Concise History of Dutch Mauritius, 1598-1710*, London: Kegan Paul International.

Morga, Antonio de (1973) 'Sucesos de las Islas Filipinas, 1609'. In Blair and Robertson, *The Philippine Islands, 1493-1898*, vol. 16 (pp. 27-209).

Muhammad Rabi, ibn Muhammad Ibrahim (1972) *The Ship of Sulaiman*, trans. John O'Kane, London: Routledge and Kegan Paul.

Mullin, Molly (1999) 'Mirrors and Windows: Sociocultural Studies of Human-Animal Relationships', *Annual Review of Anthropology*, vol. 28 (pp. 201-224).

National Archives, Public Record Office, London, DO119/1096.

[Nederlands-Indische Bladen] (1924) 'Het Veeteeltbedrijf in den Ambtskring Soembawa-Besar' [Animal husbandry in the District Sumbawa-Besar], *Nederlands-Indische Bladen voor Diergeneeskunde en Dierenteelt (NIBDD)*, vol. 36 (pp. 356-357).

Nel, J. J. (1930) 'Perdeteelt in Suid-Afrika, 1652-1752'. MA thesis: University of Stellenbosch.

Nicols, Nicholas (1973) 'Commerce of the Philippine Islands, 1759'. In Blair and Robertson, *The Philippine Islands, 1493-1898*, vol. 47 (pp. 251-284).

Noble, John (ed.) (1886) *Official Handbook of the Cape of Good Hope*, Cape Town: Soloman.

Noordijk, P., and J. van der Weijde (1856) 'Staat van den Veerstapel in de 1ᵉ en 3ᵉ Afdeeling op Java' [Situation of Livestock in the 1st and 3rd Division of Java], *Tijdschrift voor Nijverheid in Nederlandsch Indië*, 1856, vol 3 (pp. 167-201).

Noorduyn, J. (1987) *Bima en Sumbawa: Bijdragen tot de Geschiedenis van de Sultanaten Bima en Sumbawa door A. Ligtvoet en G.P. Rouffaer.* [Bima and Sumbawa: Contributions to the History of the Sultanates of Bima and Sumbawa by A.Ligtvoet and G.P. Rouffaer], Dordrecht: Foris [KITLV, Verhandelingen 129].

Noske, Barbra (1989) *Humans and Other Animals: Beyond the Boundaries of Anthropology*, Pluto Press, 1989.

Nusa Tenggara (1979) *Nusa Tenggara dalam Angka, 1978* [Nusa Tenggara in Statistics, 1978], Mataram: Kantor Sensus dan Statistik Propinsi NTB.

—— (1988) *Nusa Tenggara dalam Angka, 1987* [Nusa Tenggara in Statistics, 1987], Mataram: BAPPEDA dan Kantor Statistik Propinsi NTB.

Official Year Book of the Union, and of Basutoland, Bechuanaland Protectorate, and Swaziland (1922), Government Printer: Pretoria.

Olivier, Johannes. (1828) *Land- en Zeetogten in Nederlands Indië, en Eenige Britsche Etablissementen, Gedaan in de Jaren 1817 tot 1826* [Journeys by Land and Sea in the Netherlands-Indies, and some British Establishments, Made in the Years Between 1817 and 1826, Part II], Amsterdam: Sulpke

[On the Different Races] (1837, 1986) 'On the Different Races of the Horse in the Malayan Archipelago and Adjacent Countries'. In Moor, J.H (ed.) *Notices of the Indian Archipelago and adjacent Countries,* London: Frank Cass (pp. 189-190).

Osborne, Michael (1994) *Nature, the Exotic, and the Science of French Colonialism*, Bloomington and Indianapolis: Indiana University Press.

Overgekomen Brieven en Papieren *(OBP)* [Transmitted Letters and Papers] The Nationaal Archief in The Haque, VOC archives. OBP no. 1240; OBP no. 1246; OBP no. 1269; OBP no. 1281; OBP no. 1287; OBP no. 1294; OBP no. 1301.

Palmer, D. and Victor Buencamino (1913) 'The College of Veterinary Science, University of the Philippines', *Philippine Agricultural Review*, vol. 6 (pp. 368-370).

Pankhurst, Richard (1968) *Economic History of Ethiopia, 1800-1935*, Addis Ababa: Haile Selassie I University Press.

Parimartha, I. Gde (1995) 'Perdagangan dan Politik di Nusa Tenggara, 1815-1915', Doctoral Thesis, Vrije Universiteit, Amsterdam.

Philo, Chris and Chris Wilbert (eds) (2000) *Animal Spaces, Beastly Places: New Geographies of Human-Animal Relations*, London: Routledge.

Pigeaud, Theodore G. T. (1960-1963) *Java in the 14ᵗʰ Century: A Study in Cultural History,* 5 vols, The Hague: Nijhoff (KITLV Translation series, 4).

Playne, Somerset (1910-1911) *Cape Colony (Cape Province): Its History, Commerce, Industries and Resources*, London: Foreign and Colonial Compiling and Publishing Co.

'Pliego de Condiciones de Carrera de Caballo de la Provincia de la Pampanga' [Details of the Main Horse Track of the Province of Pampanaga] (1879) Philippine National Archives, Carreras de Caballos, Bundle 1.

'Pliego de la Matanza y Limpieza de Reses, Fecha 4 de Mayo de 1880, Manila 1 March 1889' [Folder on the Slaughtering and Cleaning of Cattle, 1889] (1889), Philippine National Archives, Matanzas y Limpieza de Reses.

Pongsripian, Winai (ed.) (1991) *Khamhaikan khun luang wat pradu songtham* [Testimony of Khun Luang Wat Pradu Songtham], Bangkok: Secretariat of the Prime Minister's Office.

—— (ed.) (2005) *Kot monthianban chabab chaloem phra kiat* [Palatine Law, Royal Celebratory Edition], Bangkok: Thailand Research Fund.

Postma, Dirk (1897)*De Trekboeren te St. Januario Humpata*, Amsterdam: [no publisher].

Postma, M. (1974) *Tales from the BaSotho*, Trans. S. McDermid. Austin: University of Texas Press.

'Proyecto de Renovación y Mejora de la Cria Caballar' [Project to Reinvigorate and Improve the Breed of Horse of this Archipelago] (1881), Philippine National Archives, Raza de Caballeria de Filipinas, Formento y Mejora.

Ptak, Roderich (1991) 'Pferde auf See, ein Vergessener Aspekt des Maritimen Chinesischen Handels im Frühen 15. Jahrhundert', *Journal of the Economic and Social History of the Orient*, vol. 34, no. 2 (pp. 199-233).

—— (1999) *China's Seabourne Trade with South and Southeast Asia (1200-1750)*, Aldershot: Ashgate.

Purchas, Samuel (1625) *Purchas His Pilgrims*, Book III, London: William Stansby.

'Quarantine Stations' (1910), *Philippine Agricultural Review*, vol. 3 (pp. 27-28).

Quiason, Serafin D. (1966) *English 'Country Trade' with the Philippines, 1644-1765*, Quezon City: University of the Philippines Press.

Raben, Remco and Dhiravat na Pombejra (eds.) (1997) *In the King's Trail*, Bangkok: Royal Netherlands Embassy.

Raber, Karen and Treva J. Tucker (eds), *The Culture of the Horse: Status, Discipline, and Identity in the Early Modern World*, New York: Palgrave Macmillan, 2005.

Radermacher, Jacob Cornelis Matthieu (1786) 'Korte Beschrijving van het Eiland Celebes en de Eilanden Floris, Sumbauwa, Lombok en Bali' [Short Description of the Island of Celebes and the Islands Flores, Sumbawa, Lombok and Bali], *Verhandelingen van het Bataviaasch Genootschap voor Kunsten en Wetenschappen (VBG)*, vol. 4 (pp. 143-196).

Raffles, Thomas Stamford (1830) *The History of Java*, 2 vols., London: Murray [1st edition 1817].

Reinwardt, Caspar Georg Carl (1858) *Reis naar het Oostelijk Gedeelte van den Indischen Archipel, in het Jaar 1821* [Journey to the Eastern Part of the Indian Archipelago, in the Year 1821], Edited by W.H. de Vriese. Amsterdam: Frederik Muller.

'Relación de los Caballos, [Pertaining to Horses], San Pascual 10 February 1889' (1889) Philippine National Archives, Carruajes, Carros y Caballos, Bundle 7.

'Remarks on the Philippine Islands, and on their Capital Manila, 1819 to 1822, By an Englishman, 1828' (1973). In Blair and Robertson, *The Philippine Islands, 1493-1898*, vol. 51 (pp. 71-181).

Report of the Philippine Commission to the President, January 31, 1900 (1900) Washington: Government Print Office.

Report of the Philippine Commission, 1903 (1904) Washington: Government Printing Office.

Report of the Philippine Commission, 1904, (1905) Washington: Government Printing Office.

Report of the Philippine Commission, 1905 (1906) Washington: Government Printing Office.

Report of the Philippine Commission, 1913 (1914) Washington: Government Printing Office.

Richards, John (2003) *The Unending Frontier: An Environmental History of the Early Modern World*, Berkeley: University of California Press.

Richardson, Clive (1998) *The Horse Breakers*, London: J.A. Allen.

Ricklefs, Merle Calvin (1981) *A History of Modern Indonesia*, Houndmills, Basingstoke, Hampshire and London: Macmillan [Macmillan Asian Histories Series].

Rimmer, Peter J. (1990) 'Hackney Carriages, Syces and Rikisha Pullers in Singapore: A Colonial Registrar's Perspective on Public Transport, 1892-1923'. In Peter J. Rimmer and Lisa M. Allen (eds), *The Underside of Malaysian History: Pullers, Prostitutes and Plantation Workers*, Singapore: Singapore University Press (pp. 129-60).

Ritvo, Harriet (1987) *The Animal Estate: The English and Other Creatures in the Victorian Age*, Cambridge: Harvard University Press.

—— (1994) 'Possessing Mother Nature: Genetic Capital in 18th-Century Britain'. In Susan Staves and John Brewer (eds), *Early Modern Conceptions of Property*, London: Routledge, 1994 (pp. 413-426).

—— (1997) *The Platypus and the Mermaid and other Figments of the Classifying Imagination*, Cambridge and London: Harvard University Press.

—— (2002) 'History and Animal Studies', *Society and Animals*, vol. 10, (pp. 403-406).

—— (2004) 'Animal Planet', *Environmental History*, vol. 9, no. 2 (pp. 204-220).

Romanes, George (1882) *Animal Intelligence*, New York: Appleton.

—— (1883) *Mental Evolution in Animals*, London: Kegan, Paul, Trench.

Romero, Patricia W. (1997) *Lamu: History, Society, and Family in an East African Port City*, Princeton: Princeton University Press.

Roorda van Eysinga, Philippus Pieter (1830-1832) *Verschillende Reizen en Lotgevallen van S. Roorda van Eysinga* [Various Travels and Adventures of S. Roorda van Eysinga], 4 vols, Amsterdam, van der Heij.

Rosenthal, Eric (1964) *Encyclopaedia of Southern Africa*, London: Warne.

Rouffaer, Gerret Pieter and Jan Willem IJzerman (eds) (1915) *De Eerste Schipvaart der Nederlanders naar Oost-Indië onder Cornelis de Houtman, 1595-1597; I: D'Eerste boeck, van Willem Lodewyckz* [The First Sea Journey of the Dutch to the East Indies under Cornelis de Houtman, 1595-1597; I: The First Book, by Willem Lodewyckz],'s-Gravenhage: Nijhoff (Werken Linschoten Vereeniging 7).

Ross, R. (1993) *Beyond the Pale, Essays on the History of Colonial South Africa.* Hanover: Wesleyan University Press.

Rothfels, Nigel (2002) *Savages and Beasts: The Birth of the Modern Zoo*, Baltimore and London: Johns Hopkins University Press.

—— (ed.) (2003) *Representing Animals*, Bloomington: University of Indiana Press.

Rowbotham, Shelia (1973) *Hidden from History: 300 Years of Women's Oppression and the Fight Against It*, London: Pluto Press.

Ruangsilp, Bhawan (2007) *Dutch East India Company Merchants at the Court of Ayutthaya: Dutch Perceptions of the Thai Kingdom, c.1604-1765*, Leiden/ Boston: Brill.

Rueda y Mendoza, Diego de (1973) 'Royal Festivities at Manila, 1 August, 1625'. In Blair and Robertson (eds), *The Philippine Islands, 1493-1898*, vol. 22 (pp. 50-61).

Rupke, Nicolas (ed.) (1987) *Vivisection in Historical Perspective*, London: Routledge.

Russell, Nicholas (1986) *Like Engend'ring Like: Heredity and Animal Breeding in Early Modern England*, London: Cambridge University Press.

Said, Edward (1978) *Orientalism*, New York: Pantheon Books.

Salahuddin, H. St. Maryam R. and H. Abdul Wahab H. Ismail (1988/1989) *Pemerintah Adat Kerajaan Bima: Struktur dan Hukum.*[Adat Prescriptions of the Kingdom of Bima: Structure and Legislation], Mataram: Departemen Pendidikan dan Kebudayaan. [Museum Negeri Nusa Tenggara Barat].

Salisbury, Joyce (1986) *The Beast Within: Animals in the Middle Ages*, New York: Routledge.

Sanders, Peter (1975) *Moshoeshoe: Chief of the Sotho*, London: Heinemann.

Sastrodihardjo, Raden Soekardjo (1956) *Beberapa Tjatatan Tentang Daerah Pulau Sumbawa* [Some Remarks Concerning the Island of Sumbawa], Singaradja: Djawatan Petanian Rakjat Propinsi Nusa Tenggara.

Savory, P. (1962) *Basuto Fireside Tales*, Cape Town: Howard Timmins.

Savory, Theodore H. (1979) *The Mule, a Historic Hybrid*, Shildon: Meadowfield Press.

Scarr, Deryck (2000) *Seychelles since 1770, History of a Slave and Post-Slavery Society*, London: Hurst & Company.

Schérer, André (1985) *La Réunion*, Paris: Presses Universitaires de France.

Schooneveld-Oosterling, J. E. (ed.) (1997) *Generale Missiven van Gouverneurs-Generaal en Raden aan Heren XVII der Verenigde Oostindische Compagnie*

[Correspondence of the Governors-general to the Board of the Dutch East India Company], Vol. XI, The Hague: Martinus Nijhoff.

Schouten, Joost (1986) 'A Description of ... Siam'. In François Caron and Joost Schouten, *A True Description of the Mighty Kingdoms of Japan and Siam* (facsimile of 1671 London edition), Bangkok: The Siam Society (pp. 121-152).

Schreuder, P. J. (1915) *'The Cape Horse'*, PhD thesis, Ithaca: Cornell University.

Scott, William (1994) *Barangay: Sixteenth Century Philippine Culture and Society*, Quezon City: Ateneo de Manila.

Selections from the Dutch Records of the Ceylon Government No. 5. (1946) Colombo: State Printing Corporation.

Serpell, James (1986) *In the Company of Animals: A Study of Human-Animal Relationships*, Oxford: Blackwell.

Sessions, H. (1903) *Two Years With Remount Commission*, London: Chapman and Hall.

Shepard, Paul (1995) *The Others: How Animals Made Us Human*, Washington, D.C.: Island Press.

Sibinga Mulder, J. (1927) 'De Economische Beteekenis van het Vee in Nederlandsch Oost-Indië en de Regeeringszorg Ervoor' [The Economic Significance of Livestock in the Netherlands East Indies and the Government Attendance for it], *De Indische Gids*, vol. 19 (part I) (pp. 308-327).

Slatta, Richard (1990) *Cowboys of the Americas*, New Haven: Yale University Press.

Smith, George Vinal (1977) *The Dutch in Seventeenth-Century Thailand*, De Kalb: University of Northern Illinois Center for Southeast Asian Studies.

Smith, Neil (1996) 'The Production of Nature'. In George Robertson, Melinda Mash, Lisa Tickner, Jon Bird, Barry Curtis, and Tim Putnam (eds), *Future Natural Nature/Science/Culture,* London and New York: Routledge (pp. 35-54).

'Sobre los Gravisimos Perjuicios, 30 May 1807' [Concerning the Severest of Wrongs] (1807), Philippine National Archives, Spanish Manila, Reel 5.

Sparrman, A. (1975-1977) *A Voyage to the Cape of Good Hope, Towards the Antarctic Polar Circle, and Around the World, But Chiefly into the Country of the Hottentots and Caffres, From the Year 1772 to 1776* [1785], Cape Town: Van Riebeeck Society.

Stampa [no first name given] (1846) 'De Paarden in Nederlandsch Oost-Indië' [The Horses in the Netherlands East Indies], *Militaire Spectator: Tijdschrift voor het Nederlandsche Leger,* vol 14 (pp. 167-174).

Stanton, William (1975) *The Great United States Exploring Expedition of 1838-1842,* Berkeley, Los Angeles and London: University of California Press.

Stibbe, David Gerhard (ed.) (1919) *Encyclopaedie van Nederlandsch-Indië*, Tweede druk, Derde deel, The Hague/Leiden: Martinus Nijhoff/E.J. Brill.

Straits Settlements Annual Reports, Singapore: Government Printing Office.

Summerhays, R. S. (1954) *The Observer's Book of Horses and Ponies*, London: Frederick Warne & Co.

'Supresión del Impuesto sobre los Caballos, Abra 1889' [Abolition of the Tax on Horses, Abra 1889] (1889), Philippine National Archives, Carruajes, Carros y Caballos, Bundle 7.

'Surra' (1908) Philippine Agricultural Review, vol. 1, (pp. 119-121).

Swabe, Joanna (1999) *Animals, Disease and Human Society: Human-Animal Relations and the Rise of Veterinary Medicine*, London: Routledge.

Swart, Sandra (2003) 'Riding High - Horses, Power and Settler Society, c. 1654-1840', *Kronos*, vol. 29, (Environmental History Special Issue), November (pp. 47-63).

—— (2004) '"Race" Horses - A Discussion of Horses and Social Dynamics in Post-Apartheid Southern Africa'. In N. Distiller and M. Steyn (eds), *Under Construction: 'Race' and Identity in South Africa Today*, Sandton: Heinemann, pp. 13-24

—— (2005) '"Horses! Give Me More Horses!" - White Settler Society and the Role of Horses in the Making of Early Modern South Africa'. In Karen Raber and Treva Tucker (eds), *The Culture of the Horse: Status, Discipline, and Identity in the Early Modern World*, Basingstoke: Palgrave Macmillan, pp. 311-328.

Syaraswati and M. Yusuf H. Umar (1985/1986) *Upacara U'a Pua di Kabupaten Bima (Pengaruh Agama Islam)* [The U'a. Pua Ceremony in Kabupaten Bima (the Influence of the Religion of Islam)], Mataram: Departemen Pendidikan dan Kebudayaan, Museum Negeri Nusa Tenggara Barat.

'Tarifa de los Honorarios' [Price of Services] (1864), Philippine National Archives, Veterinarios.

Taylor, Jean Gelman (2003) *Indonesia: Peoples and Histories*, New Haven/London: Yale University Press.

Taylor, Pamela York (1994) *Beasts, Birds, and Blossoms in Thai Art*, Kuala Lumpur: Oxford University Press.

Teenstra, Marten Douwes (1828-1830) *De Vruchten mijner Werkzaamheden, Gedurende mijne Reize naar Java* [The Fruits of my Labours, During my Journey to Java], 3 vols, Groningen: Eekhoff.

Temple, Richard Carnac (ed.) (1919) *The Travels of Peter Mundy in Europe and Asia, 1608-1667, vol. III, part 1 (1634-1637)*, London: Hakluyt Society (Works issued by the Hakluyt Society, 2nd series, 45).

'Testimonio 22 June 1849' [Testimoy of 22 June 1849] (1849), Philippine National Archives Matanzas y Limpieza de Reses.

Theal, George M. (1897-1905) *Records of the Cape Colony*, vol. XIV, London: Clowes.

Thirsk, Joan (1978) *Horses in Early Modern England For Service, For Pleasure, For Power*, Reading: University of Reading.

—— (ed.) (1985) *The Agrarian History of England and Wales, Volume VII, 1640-1750: Agrarian Change*, Cambridge: Cambridge University Press.

Thomas, Keith (1983) *Man and the Natural World,* New York: Pantheon Books.

Thompson, George (1827) *Travels and Adventures in Southern Africa,* London: Henry Colburn.

Thompson, L. (1990) *A History of South Africa,* New Haven: Yale University Press.

Thorn, William (1993) *Memoir of the Conquest of Java: With the Subsequent Operations of the British Forces in the Oriental Archipelago.* Singapore: Periplus Editions [Originally 1815].

Thorndike, Edward (1898) 'Animal Intelligence: An Experimental Study of the Associative Processes in Animals', *Psychological Monographs,* vol. 2, no. 4 and 8.

Thorndike, Edward (1911) *Animal Intelligence: Experimental Studies,* New York: Macmillan.

Thornton, Russel William (1936) *The Origin and History of the Basotho Pony,* Morija: Morija Printing Works.

Thunberg, Charles (1793) *Travels in Europe, Africa and Asia, Performed in the Years 1770 and 1779,* vol. II, London, 1793.

Toth, D. (2000) 'The Psychology of Women and Horses'. In GaWaNí PonyBoy (ed.), *Of Women and Horses,* Irvine: BowTie.

Tuan, Yi-Fu (1984) *Dominance and Affection: The Making of Pets,* New Haven: Yale University Press.

Tuchman, B. (1966) *The Proud Tower,* Macmillan: New York.

Tylden, G. (1950) *The Rise of the Basuto,* Cape Town: Juta.

—— (1980) *Horses and Saddlery, An Account of the Animals used by the British and Commonwealth Armies from the Seventeenth Century to the Present Day, with a Description of their Equipment,* London: J. A. Allen.

'Una Epizootia en Filipinas' [An Epizootic in the Philippines] (1888) Philippine National Archives, Veterinarios.

UNDP/FAO (1982) *National Conservation Plan for Indonesia. Volume IV: Nusa Tenggara.* Bogor: UNDP/FAO. [Field Report no. 44 of The National Parks Development Project, based on the work of John MacKinnon and others].

Valentijn, François (1724-1726) *Oud en Nieuw Oost-Indiën* [Old and New East Indies], 5 vols. in 8 parts. Amsterdam: Onder den Linden.

Van der Merwe, F. J. (1981) *Ken Ons Perderasse* [Know our Horse Breeds], Cape Town: Human and Rousseau.

Van der Walt, A. J. H. (1928) *Die Ausdehnung der Kolonie am Kap der Guten Hoffnung (1700-1779): Eine Historisch-Ökonomische Untersuchung über das Werden und Wesen des Pionierlebens im 18. Jahrhundert,* Berlin: Ebering.

Van Riebeeck, J. (1660) *Journal of Jan Van Riebeeck,* vol. III, 1954, 14 December.

Van Riebeeck, Jan H. and B. Thom (ed.) (1954) *Journal of Jan Van Riebeeck.* vol. II, 1656-1658, Cape Town: A.A. Balkema.

Van Sittert, Lance and Sandra Swart (eds.) (2003) 'Canis Familiaris - A Dog History of South Africa', *South African Historical Journal,* vol. 48, (pp. 138-251).

Van Spaan, Gerrit *De Gelukzoeker,* Rotterdam: De Vries, 1752, p. 28; J. van Riebeeck, *Journal of Jan Van Riebeeck,* vol. III, 1954, 14 December 1660.

Vasconcellos, Ernesto J. de Carvalho e (1896) *As Colónias Portuguesas*, Lisbon: Companhia Nacional Editora.

Verheijen, Jilis A.J. (1967) *Kamus Manggarai I: Manggarai-Indonesia* [A Dictionary of Manggarai I: Manggarai-Indonesian], ´s-Gravenhage: Martinus Nijhoff [KITLV].

Veth, Pieter Johannes (1894) *Het Paard onder de Volken van het Maleische Ras* [The Horse among the People of the Malay Race], Leiden: Brill.

Viana, Francisco de (1973) Memorial of 1765. In Blair and Robertson, *The Philippine Islands, 1493-1898*, vol. 48 (pp. 197–338).

Villaldea, Maria Isable Piquerras (2002) *Las Comunicaciones en Filipinas durante el Siglo XIX Caminos, Carreteras y Puentes* [Communications in the Philippines During the 19th Century Roads, Highways and Bridges], Madrid: Archiviana.

Viveiros de Castro, Eduardo (1992) *From the Enemy's Point of View: Humanity and Divinity in an Amazonian Society*, Chicago: University of Chicago Press.

Vivít, Estevan (no date) *Reseña Estadística de las Islas Filipinas en 1845* [Current Statistics of the Philippine Islands in 1845], Barcelona: Imprenta de A. Brosi.

Vliet, Jeremias van (2005) 'Description of the Kingdom of Siam', trans. L. F. van Ravenswaay. In Chris Baker, Dhiravat na Pombejra, Alfons van der Kraan and David K. Wyatt (eds.), *Van Vliet's Siam*, Chiang Mai: Silkworm.

[Volkstelling 1930] (1936) *Volkstelling 1930. Deel V: Inheemse Bevolking van Borneo, Celebes, de Kleine Soenda Eilanden en de Molukken* [Census 1930. Part V: Indigenous Peoples of Borneo, Celebes, the Lesser Sunda Islands and the Moluccas], Batavia: Landsdrukkerij.

Wallace, Robert (1896) *Farming Industries of Cape Colony*, London: King.

Ware, Vron (1992) *Beyond the Pale: White Women, Racism and History*, London and New York: Verso.

Water, W. G. C. toe (1844) 'Sampela, Een Tafereel van Bimanesche Zeden, Gewoonten en Karakters' [Sampela, a Scene of Bimanese Manners, Customs and Characters], *Tijdschrift van Nederlandsch-Indië (TNI)*, vol 6 (pp. 400–414), (pp. 549–563).

Watney, Marylian and Sanders (1975) *Horse Power*, London: Hamlyn.

Weide, J. van der, 'Iets over de op Java Voorkomende Paarden' [Data on the Horses of Java], *Tijdschrift voor Nijverheid in Nederlandsch-Indië*, 1860, vol. 7 (pp. 388–401).

Wernstedt, Frederick and J. E. Spencer (1978) *The Philippine Island World: A Physical, Cultural and Regional Geography*, Berkeley: University of California Press.

Wharton, William James Lloyd (ed.) (1893) *Captain Cook's Journal during his First Voyage round the World made in H.M. Bark ' Endeavour' 1768-71*, London: Stock.

White, Hayden (1987) *The Content of the Form: Narrative Discourse and Historical Representation*, Baltimore: Johns Hopkins University Press.

White, Lynn (1967) 'The Historical Roots of Our Ecological Crisis', *Science*, vol. 155 (pp. 1203-1207).

Wilkes, Charles (1973) *Narrative of the United States Exploring Expedition during the Years 1838, 1839, 1840, 1841, 1842.* In Blair and Robertson, *The Philippine Islands, 1493-1898*, vol. 43 (pp. 128-192).

Willerslev, Rane (2004) 'Not Animal, Not *Not*-Animal: Hunting, Imitation and Empathetic Knowledge among the Siberian Yukaghirs', *Journal of the Royal Anthropological Institute*, vol. 10 (pp. 629-652).

Williams, Bernard (1991) 'Prologue: Making Sense of Humanity'. In James Sheehan and Morton Sosna (ed.), *The Boundaries of Humanity: Humans, Animals, Machines*, Berkeley and Los Angeles: University of California Press (pp. 13-23).

Willis, Roy (1967) *Man and Beast*, New York: Basic Books, 1974

—— (ed.) (1990), *Signifying Animals: Human Meaning in the Natural World*, London: Routledge.

Wolch, Jennifer and Jody Emel (eds.) (1998), *Animal Geographies: Place, Politics and Identity in the Nature-Culture Borderlands*, London: Verso.

Wolf, Eric (1997) *Europe and the People Without History.* Berkeley: University of California Press.

Wolters, Oliver W. (1967) *Early Indonesian Commerce: A Study of the Origins of Srivijaya*, Ithaca, New York: Cornell University Press.

Worsfold, W. B. (1893) *A Visit to Java, with an Account of the Founding of Singapore*, London: Bentley and Son.

Wyatt, David K. (1884) *Thailand: A Short History*, London and Bangkok: Yale University Press.

Wyndham, H. A. (1924) *The Early History of the Thoroughbred Horse in South Africa*, London: Humphrey Milford/Oxford University Press.

Yarwood, Alexander T. (1989) *Walers: Australian Horses Abroad*, Melbourne: Melbourne University Press.

Yarwood, R. N. Evans, and J. Higginbottom (1997) 'The Contemporary Geography of Indigenous Irish Livestock', *Irish Geography*, vol. 30 (pp. 17-30).

Yates, R., C. Powell, P. Beirne (2001) *'Horse Maiming in the English Countryside: Moral Panic, Human Deviance, and the Social Construction of Victimhood', Society & Animals*, vol. 9, no. 1, (pp. 1-23).

Yearbook of Food and Agricultural Statistics (1947) Washington: Food and Agriculture Organization.

Yncidente sobre la Detención de un Caballo' [Incident to do with Detaining a Horse] (1884) Philippine National Archives, Animales Sueltos, Corregimiento de Manila.

Young, Robert (1995) *Colonial Desire: Hybridity in Theory, Culture and Race*, London: Routledge.

Zinsser, Hans (1963) *Rats, Lice, and History: Being a Study in Biography, which, after Twelve Preliminary Chapters Indispensable for the Preparation of the*

Lay Reader, Deals with the Life History of Typhus Fever, London: Paper-mac.

Zollinger, Heinrich (1850) 'Verslag van eene Reis naar Bima en Soembawa en naar Eenige Plaatsen op Celebes, Saleyer, en Floris Gedurende de Maanden Mei tot December 1847' [Report of a Journey to Bima and Sumbawa and to Several Places on Celebes, Saleyer and Flores during the Months May to December 1847], *Verhandelingen van het Bataviaasch Genootschap voor Kunsten en Wetenschappen* (VBG), vol. 23 (pp. 1-224).

Index

Abra, 89

Abra River, 95

accidents and injuries to horses, 95-96, 97

acclimatisation, 111, 112, 113, 119, 120, 202 nn.43, 50, 204 n.68

Aceh, 35-36, 37, 48, 69, 172 n.13

adornment of horses, 68, 118-119, 187 n.13, 204-205 n.76

Adriani, Nicolaus, 53

Africa, 1, 52
　East, 27
　First World War campaigns, 26
　introduction of horses, 123
　North, 2, 14, 23, 123
　West, 9, 14, 123, 128, 206 n.1, 209 n.3, 212 n.83
　See also South Africa; Southern Africa

African Horse Sickness, 21, 25, 123, 126, 129, 130, 210 n.54

African Turf Club, 135, 136, 137

Agam breed, 37

Agra, 43

agriculture
　ladang, 61, 62
　use of horses in, 1, 13, 14, 22, 36, 57, 123, 125, 133, 207 n.12

Ajoebar, 61

Akbar, 43

Alaungpaya, King, 70

Algeria, 28

Amangkurat II, Susuhunan, 74, 77

ambling horses, 131, 132-133

Ambon, 40

Americas, 120
　dissemination of equine genes and phenotypes from, 13, 14, 23, 123, 154
　extinction of horses in, 155 n.3
　Spanish conquest of, 1, 9
　Spanish reintroduction of horses to, 32, 94, 131, 155 n.3
　transfer of species between Europe and, 9
　See also South America; United States

Andalusian horses, 85, 108

Anderson, John, 36

Anderson, Kay, 5

Anderson, Warwick, 121

Anglo-Boer War, 26, 93, 144

Angola, 26

animal rights movement, 4, 157-158 n.21

animal studies, 2-5, 156, n.9

animals
　anthropogenic use of, 3
　as 'other', 6-10, 153, 161 n.53
　centrality to human lives, 3
　consciousness (self-awareness) of, 6-7, 159 n.39, 160 n.47
　environmental effects of, 9
　in history, 2-5, 218 n.3
　in non-Western settings, 8-9, 161-162 n.56
　intelligence of, 7, 159 n.39
　psychology of, 7

anthrax, 62, 63, 97
anthropomorphism, 6, 7, 134
Arab horses, 24, 31, 34, 39, 44, 47, 59, 65, 70, 127, 135
   bred with Basutho Pony, 145, 147, 148
   influence on Cape Horse, 14, 131, 132
   influence on Javanese horse, 66, 67
   Philippines, 85, 108, 109, 113, 115, 116, 119
Arab traders, 29, 30-31, 52, 58, 59, 127, 187 n.14
Arabia, 34, 39
Argentina, 28
aristocracy, 127-128
   Priangan, 39-40
   See also elite
Arnold, David, 111
Asahan, 36
Asia, 2, 9
   Central, 1
   dissemination of equine genes and phenotypes from, 13, 14, 52, 123, 154
   East, 23-24, 29
   Inner, 22, 23, 24, 29
   West, 1
   See also Southeast Asia; Southwest Asia; and specific Asian countries
Australasia, 1, 120
Australia
   breeding in, 23
   colonial hegemonic identity, 5
   export of equids, 21, 23, 24, 25, 26, 27, 29, 31, 32, 44, 59, 108, 115
   imports of horses, 23, 120, 149
   stud stock from, 108, 115
Ayutthaya, court of
   Javanese horses for 65, 71, 72-81
   significance of horses to, 68-71
   ties with Javanese court, 78

titles of people sent to buy horses, 75-76, 189 n.52
   VOC loans to buy horses, 70, 73, 75, 76, 77, 78, 80, 81, 185

Bahrain, 27
Baker, Steve, 4
Bakewell, Robert, 11-12
Bali, 30, 43, 44, 45, 58, 65
Bandung, 44
Bangka, 72
Bangkok, 71
Bankoff, Greg, 1-18, 85-103, 105-121, 153-154
Banten, 38-39, 43, 72, 74
Barbosa, Duarte, 54
Barbs, 127, 131, 135
Barnard, Lady Anne, 132, 136
BaSotho people, 141, 143, 144, 150
Basotho Pony, 141, 142, 144-145
   inventing, 145-149
Basotho Pony Project, 216 n.25
Basutoland. *See* Lesotho
Batak highlands, 35, 45-46, 48, 172 n.17
Batak horses/ponies, 35, 36, 37, 172-173 nn.13, 18
Batavia, 57
   hub of Asian inter-exchange of horse genes, 33, 47
   imports of horses, 59
   post-road to eastern Java, 44
   Siamese horse trading in, 44, 72, 73, 76, 78, 80
   Sumbawan horses in, 57
   See also Dutch United East Indies Company (VOC)
Bataviase Ommelanden, 44
Batubara, 36, 172, n.13
Beaulieu, Augustin de, 35
Bengal, 79
Berger, John, 10
Besuki, Residency of, 42, 59, 66

Bhamo, 29

Bicol region, 31, 93, 96, 99, 100

Bicol River, 95

Bima, 43, 44, 47, 52, 54, 55-57, 58, 59, 60, 64, 171 n.2

Bima horses, 34, 39, 41, 45, 53, 54, 55-56, 57, 58

Bird, William, 136-137

blacksmiths, 98

Blambangan, 40, 42, 43

Blok, Gisela, 8

Blom, Wijbrand, 79

'Boerperd', 132, 147, 148

Bogor (Buitenzorg), 30, 44

Bombay, 24

Bondowoso, 42

Bone, Kingdom of, 55

Bonthain, 58

Boomgaard, Peter, 4, 9, 33-48

Borneo, 34, 57, 85

Borommakot, King 65, 71, 80

Bouman, M.A., 55

Bourbon (Réunion), 21, 27, 58

bovids, 22

Braud, Gabriel, 77

breed, definition of, 1, 10-11, 33, 146-147

breeding, 1, 2, 10-12
    Aceh, 35, 37, 48
    as incentive to settlement of particular areas, 48
    Batak highlands, 35, 37, 48
    Cape, 129, 130, 131-132, 133, 137
    cross-breeding, 2, 67, 72, 85, 105, 108, 114-115, 131
    in-breeding, 112-113
    Indonesia, eastern, 45-47
    Java, 38-44
    Lesotho, 144-145, 149, 216 n.25
    line-breeding, 112-113
    Minangkabau area, 37
    out-breeding, 17, 112, 113, 120

Philippines, 85, 108-111, 112-116, 119, 120, 199-202
    population density, impact of, 37-38, 40, 41, 45-46, 47, 48
    reinteelt, 59-60, 115
    royal courts, impact of, 13, 36, 37, 39, 40, 41, 42, 46, 47, 48, 54-56, 172 n.17
    scientific, 13, 111-112, 120, 150
    selective, 112
    'semi-wild' method, 60
    South Africa, 147-148, 149
    Spain, 108
    Sumatra, 35-38, 46, 47, 48
    Sumbawa, 53, 53-54, 56, 59-60, 61, 63
    VOC, impact of, 33-34, 40, 43-44, 47, 57, 79
    See also pasture and breeding; stud farms

breeds. *See* Arab horses; Andalusian horses; Basotho Pony; Batak horses/ponies; Bima horses; Cape Horse; 'Capers'; Gayo horses; Java ponies; Kedu horses; Kuningan horses; Minangkabau horses; Mongolian horses; Persian horses; Philippine Horse; Priangan horses, 39; Sembrani horses; Tambora horses; thoroughbred horses; Timor pony; Walers

Britain, 5, 108
    imports of horses to India, 24-25, 31

British East Asia Company, 178 n.13

Buddha, 65, 186 n.1

buffaloes, 35, 36, 40, 57

Buffon, Georges, 111, 117

Buitenzorg (Bogor), 30, 44

Bulliet, Richard, 9

bullocks, 86

Burchell, William, 135, 137, 209 n.39

Burias, 91

Burma, 23, 26, 34, 35, 70
Bushman (Khoisan; San), 124, 125,
    127, 128, 129, 142, 143, 206-207
    n.6

Calcutta, 24
Cambodia, 70, 119
Campostela, 89
Cape, South Africa 21, 25, 26, 143
    attacks on horses by indigenous
    people, 128, 129, 139, 209-210
    n.46
    breeding, 129, 130, 131-132, 133,
    137
    established as refreshment station,
    14, 124-126
    European settlement, 123,
    124-125, 127, 128, 139
    export of horses, 24, 131-132, 133
    feral horses, 124, 130
    growth of horse trade, 129-131,
    139
    horse population, 130
    introduction of horses, 14, 23,
    124-125, 127-128, 139
    racing in, 131-132, 134-138, 139
Cape Horse, 15, 124, 130, 131-134,
    135, 139, 144, 147, 148. *See also*
    Capers
Cape Verde Islands, 23
'Capers', 24, 25, 132. *See also* Cape
    Horse
carabaos, 86, 92, 93, 97, 98, 101
carriages and carts, 29, 57, 60, 64,
    86-87, 89, 90, 96, 191 n.7
Casey, H., 116
castration, 59, 115-116, 203 n.60
Catanduanes Islands, 116
Cebu, 89, 93, 99
Celebes (Sulawesi), 34, 47, 55, 57, 60
celebrity horses, 55, 134, 213 nn.86,
    89

ceremonial use of horses, 13, 55, 71.
    See also royal processions
Ceylon, 24, 34, 57, 79
Chao Phra Khwan, 70
China, 1, 12, 22, 78, 105
    breeding of equids, 29
    export of equids, 31, 52, 76, 85
    imports of equids, 23-24, 29
China Sea, 32
Choisy, Abbé de, 69
Cianjur, 40
Cirebon, 38, 39, 43, 47, 48, 72, 78, 80
Clarence-Smith, William G., 21-32
class, and racing, 136-138, 139
Clutton-Brock, J., 11
colonisation. *See* imperialism
colour of horses, 45
    Java, 78
    Siam, 71, 75, 77
    Sumbawa, 54
Couperus, G.W., 53
courts. *See* royal courts
Crawfurd, John, 53, 68, 178 n.13
Crosby, Alfred, 1, 2, 9, 120, 150, 161
    n.54
Cullion Island, Palawan, 108
culture
    influence of animals on, 2-3
    See also nature–culture distinction;
    popular culture

Daendels, Governor-General, 44
Dalen, Adriaan van, 57
Dampier, William, 35, 85, 102
dancing horses, 67, 77
dari, 54-55
Darwin, Charles, 7, 112
Davao, 89
Davenport Adams, W.H., 53
Dawkins, Marian, 7
De Jong Boers, Bernice, 51-64, 101
decoration of horses, 68, 118-119,
    187 n.13, 204-205 n.76

deer hunting, 36, 39, 57, 62

deforestation, 37, 41, 48, 101

Delhi, 24, 43

Deli, 29, 35, 36

'Deli ponies', 29

Descartes, René, 159, n.39

Deshima, 34

Dhiravat na Pombejra 65-81

Diamond, Jared, 11

dietary regime of horses, 12, 101-102, 109, 196 n.56, 209 n.41. *See also* pasture

diplomatic networks, 1-2

Diponegoro, Prince, 56, 180-181 n.35

diseases of horses, 13, 21, 22, 25, 27, 32, 61-62, 63
  Philippines, 92-95, 97-100, 194 n.32

diseases of humans, and tropics, 111, 116, 117-118, 120

distemper, 97, 129

Dobbin, Christine, 37

domestication, 1, 6, 11, 120, 154

Dompo, 45, 52

donkeys, 22, 23, 26, 27

draught horses, 17, 29, 57, 60, 86, 106, 113, 124-125, 127, 128, 131, 133, 135, 137, 139, 141. *See also* pack horses; transport and haulage, use of horses for

Duckitt, William, 133, 136

Duff Gordon, Lucie, Lady, 133, 135-136

Dutch United East Indies Company (VOC), 13, 123
  and Cape, 14, 23, 124, 125, 127, 128, 130, 132, 134, 139, 208 n.22
  and Siamese horse-buying expeditions 65, 71, 72-81
  Batavias Uitgaand Briefboek, 71
  Generale Missiven, 71
  influence on breeding, 33-34, 40, 43-44, 47, 57, 79

Javano-Siamese diplomatic alliance against, 74
  Overgekomen Brieven en Papieren, 71
  ships, 34, 72, 73-74, 75-76, 77, 79, 80, 134
  trade with Java, 38-39, 40-41, 43-45
  trade with Sumbawa, 46, 57, 127, 209 n.35
  trade with Timor, 46

East India Company (English), 24, 74, 135

East Indies. *See* Netherlands East Indies

Elder, Glen, 128

elephants, 9, 34, 35, 38, 42, 43, 65, 76
  as gifts, 77, 79, 119
  fights involving, 69
  military use, 69, 70
  processional use, 68, 71, 75
  transport, use for, 29

elite, association of horses with, 39-40, 127-128, 136-138, 139, 209 n.42. *See also* aristocracy; royal courts

Emel, Jody, 3, 4, 5, 128

empire. *See* imperialism

English horses. *See* thoroughbred horses: English

environmental history, 2, 150

environmental impact of animals, 9-10

environmental impact of horses
  Philippines, 100, 101-103
  Sumbawa, 62-63

Ethiopia, 23, 27

Europe, 1, 2, 8-9
  dissemination of equine genes and phenotypes from, 13, 14, 23, 39, 123, 154
  transfer of species between Americas and, 9

See also neo-Europe

Fatehpur Sikri, 43
feral horses
    Cape, 124, 130
    Philippines, 95, 101, 102–103
First World War, 26, 31, 32
food, horses as, 22, 57, 58, 97, 100,
    194–195 n.40
France, 28, 74
Francis, Emanuel, 58, 182 n.54
Freijss, J.P., 56, 58
Fudge, Erica, 6, 154

gaited horses, 211 n.62
Gamron, 34
Gayo area, 48
Gayo horses, 172, n.13
geographies, animal, 2
gerobak, 59
Gervaise, Nicolas, 68
gifts, horses as, 33, 45, 63, 67, 70, 72,
    74, 76–77, 78–79, 119, 141, 205
    n.79
glanders, 62, 63, 97, 185 n.82
Goa, 55, 110
Goens, Rycklof van, 41
grasslands. *See* pasture
Gresik, 40, 43
greyhound breeding, 11
Groeneveld, W. 35, 40, 52
Guam, 31
Guillemard, F.H.H., 54
Gunung Api (Sangeang), 53, 54, 55,
    56, 101

Hackneys, 130, 211 n.59, 212 n.81
Haggard, H. Rider, 133
'hand', measurement of horses, 107,
    164 n.89, 198–199 n.13, 208 n.32
Haraway, Donna, 8
haulage. *See* transport and haulage,
    use of horses for

Havelock, Sir Henry, 24
hegemony, culture of, 6, 127, 149
hemorrhagic septicaemia, 63
Hindus, 22
    sacred cows, 159, n.34
historiography
    and human–animal relationship, 8,
    150
    and women's history, 8
    imperial, 153–154
    interactions between people,
    animals and land, 154, 218 n.4
history of horses, 154, 206 n.3
Hodson, George, 24
holy horse, 55
Horn of Africa, 26, 27
horse carriages. *See* carriages/carts
horse guards. *See* royal horse guards
horse-racing. *See* racing
horse-riding. *See* riding
horse-stirrup, 153, 217 n.1
'horse-whims', 26
'Hottentots' (Khoikhoi), 126, 127, 143,
    206–207 n.6
human–animal interactions, 2, 8, 154
    and breeding, 12
    Philippines, 86
    purposes of studying, 4
    threat of human practices to
    animals, 3
    See also pet-keeping relations
hunting, use of horses in, 22, 36, 39,
    40, 41, 42, 57, 62, 63, 89
Hurtado de Corcuera, Sebastian, 119
hybridity, 148

identity construction
    animals in, 2, 4–5, 128
    Cape, horses in, 123–124, 130–131
    colonial culture, horses in, 131
    indigenous identity, animals as
    signifiers, 217 n.39
    Lesotho, ponies in, 144, 149, 150

Peruvian Paso in, 131
Southern Africa, horses integral to, 14
Spanish, and the Philippine Horse, 116-119, 121
Ilhas de Cavalhos, 34
Ilocos region, 31, 93, 96, 99, 100
imperialism 1-2
  and alteration of European societies, 153
  and dissemination of genes and phenotypes, 13, 14, 23, 123, 154
  and indigenous people, 121, 153-154
  horses as agents of, 1-2, 149-150
  relationship between environment, science and, 120
India, 1, 14, 21, 23, 34, 67, 74, 132
  breeding of equids, 24, 29, 70
  export of horses, 52, 76, 93, 108, 109, 110, 113, 119
  imports of equids, 23, 24-25, 26, 27, 29, 31
  Southeastern, 13
  stud stock from, 108, 109, 110, 113, 119
  See also specific cities
'Indian Mutiny', 24
Indian Ocean, maritime horse trade, 13, 15, 21-32, 149
indigenous people, 2, 118
  attacks on horses, 128, 129, 139, 209-210 n.46
  Cape, 14, 123-124, 125, 126, 127, 128, 129, 133, 139
  Java, 38
  Philippines, 98, 107, 114, 115-116, 117, 119, 121, 205 n.78
  South Africa, 23
  See also BaSotho people; Indios; Khoikhoi; Khoisan; Mestizos; Sotho
Indios, 114, 119, 205 n.78
Indonesia, 13, 33-48

Eastern, 45-47
export of equids, 21, 23, 26, 28, 29, 47-48
imports of equids, 31, 47-48
movement of horses by sea, 22
shifts in horse producing areas, 48
See also Java; Sumatra; Sumbawa
Industrial Revolution, 8, 112
internal combustion engine, and need for horses, 22, 32, 48, 60, 63
introduced species, 9-10
'invention' of the horse, 10-15, 111-116, 145-149
Irrawaddy River, 29
irruptive oscillation, 10
Islam, 9, 57, 177, n.5

Jaffna, 34, 44, 79
Jahangir, 43
Jambi, Sultan of, 72
Japan, 24, 29, 34, 85, 105
Jara La Manggila, 55, 179-180 n.27
Jasper, J.E., 55
Java, 13, 14, 22, 23, 30, 31, 34, 57, 93
  breeding in, 38-44, 46, 47-48
  cattle line between Lesser Sunda Islands and, 59
  Central, 42, 43, 56
  Eastern, 42-43, 44, 72, 78, 80
  export of horses, 44-45, 65-81, 127, 130
  horse/people ratio, 43-44, 48
  imports of horses, 38-39, 42-43, 44, 45, 46, 52, 54, 57, 58, 59
  literature on horses, 34
  Western, 39
  women dancers wanted by Siamese court, 77-78
  See also specific placenames
'Java ponies', 29, 66, 130, 131
Java War, 37, 66, 180-181 n.35
'Javanese horse' (in VOC writings), 65
Jember, 59

Jene Jara, 55
Jepara, 38, 44, 78
Jortan, 43
jousting tournaments, 36–37, 41, 43, 69
Jovantes, Felipe, 98

Kabikolan, 94
Kaempfer, Engelbert, 69
Kalinjamat, 78
Kandy, 34
Kangga, 56
Kapitang, 55
Kartasura, 44, 78, 79
Kedu area, 41, 42, 45–46, 47, 48
Kedu horses, 30, 39, 41, 42, 44, 47
Keeling, William, 187 n.13
Kennedy, John, 160, n.47
Kenya, 28
Khoikhoi ('Hottentots'), 126, 127, 143, 206–207 n.6
Khoisan (Bushman; San), 124, 125, 127, 128, 129, 142, 143, 206–207 n.6
King, Dick, 134, 213 n.87
Knight, John, 9
Kolbe, Peter, 126
kommando, 128, 139, 210 n.49
Koninklijke Paketvaart Maatschappij, 30, 59
Korea, 23
Kosathibodi (Pan), Chaophraya, 75
Kuda Sembrani, 38
Kulis, 59
Kuningan horses, 38, 39, 42, 44
Kuningan upland valley, 39, 45–46
Khunluang Wat Pradusongtham, 70
Kuperus, Gerrit, 57, 181 n.43

La Loubère, Simon de, 68, 69, 72, 73
Laambu, 56
Laguna, 87
Lamarck, Jean Baptiste, 111, 112

Landry, Donna, 12
Laos, 68
Lategan, A.W., 147
Law, Robin, 9
Le Roux, C.C.F.M., 52–53
Leach, Edmund, 159 n.35
leisure pursuits involving horses, 22, 149. *See also* hunting; polo, racing; riding
Leopold, Aldo, 218 n.4
Lepanto, 89
Lesotho, 141–150, 215 n.1
    breeding, 144–145, 149, 216 n.25
    export of horses, 144
    imports of horses, 143–144
    horse as symbol of masculine identity, 142, 150, 215 n.4
Lesser Sunda Islands, 28, 30–31, 34, 45, 46, 47, 48, 59, 62, 176
Lévi-Strauss, Claude, 2, 5
Leyte and Samar, 93, 99, 100
Lichtenstein, Heinrich, 132
Licios, Mt, 89
Ligtvoet, Albertus, 56, 181 n.39
Locke, John, 12
Lombok, 30, 43, 45, 58, 59, 65
Lucknow, 24
Lush, J., *The Genetics of Populations*, 11
Luzon, 85, 86, 93, 94, 95, 96, 99, 101, 102, 114

Macartney, Lord, 136
Macau, 29
Madagascar, 26, 27
Madras, 24, 131
Madura, 42, 43, 62
Majapahit, 52
Makassar, 34, 39, 44, 45, 46–47
Malakka, 58, 85
Malang, 42
Malaya, 22, 29, 31, 34
Mandailing horse, 172–173 n.18

Mandela, Nelson, 141, 215 n.2
manes, cutting of, 70
Mangkubumi, Sultan, 69
Manila, 31-32, 76, 86, 89-91, 92, 96-97, 99, 100, 110, 118
Manila Jockey Club, 91, 92
Manipuri horsemen, 70
manure, horse, 41
Marche, Alfred, 97
'Marengo', Napoleon's horse, 134, 213 n.89
mares, 212 n.83
    Cape, 133-134
    Indonesia, 59, 60, 91
    Java, 73, 79
    Philippines, 17, 91, 108, 109, 110, 112, 113, 115, 120, 200-201 n.33
Marsden, William, 36
Martaban, 70
Mascarene Islands, 21, 23. *See also* Mauritius
    plantation markets in, 26-28
masculine identity, 5, 142, 150, 215 n.4
Massawa, 28
Mataram, 40, 41, 43, 44, 74, 77, 78, 79
Maulud, 55
Mauritius, 21, 24, 27, 58, 93, 130
Meerhoff, Pieter, 126
Melville, Elinor, 9
Mestizos, 31, 119, 205 n.78
Mexico, 9, 85, 119, 155, n.3
military use of equids, 2, 22, 149
    Anglo-Boer War, 26
    Cape, 14, 123, 129, 139
    First World War, 26, 31, 32
    Hispanic society, 86
    horses as heroes, 56, 134
    horses as symbols of supremacy, 119
    India, 24, 25, 187 n.21
    Java, 30, 31, 40-41, 56
    Lesotho, 143-144

Philippines, 109, 110, 119, 205 n.80
Siam, 68, 69, 70, 75
Southeast Asia, 13, 65
Southern Africa, 14, 25, 26, 128, 134
Sumatra, 36
Sumbawa, 57, 63
VOC, 40-41
Minangkabau highlands, 37, 45
Minangkabau horse, 37, 38, 40, 172-173 n.18
Mindanao, 13, 85, 89, 93, 99, 101
Mindoro, 31, 93, 94, 96, 99, 101
mining, use of equids in, 26
Mokka, 34
Moluccas, 34, 40, 45
Mongolian horses, 14, 35, 127
'Moors', 42, 68, 187 n.14. *See also* Arab traders; Muslims
Moshoeshoe II, King, 141, 143-144, 215 n.2, 216 n.9
motorised transport, and need for horses, 22, 32, 48, 60, 63
Moyo, 56
Mozambique, 26
Mughal horse guards, 68, 72
Muḥammad Rabi ibn Muḥammad Ibrahim, 69-70
mules, 22, 23, 24, 26, 27, 32, 89, 109, 114, 130, 203 n.57, 214 n.104
Mundy, Peter, 35
Muslims, 29, 42, 55, 76
mustangs, 85, 211 n.60

naga, 65, 186 n.1
Nagana, 21, 26, 27
Nagarakertagama, 43
Narai, King 65, 68, 72-74
Naresuan, King, 70
Natal, 25
nature, belief in human right to dominion over, 111-112, 113, 120

nature-culture distinction, 3, 4, 5, 8,
12, 139
neo-Europes, 10, 120, 161, n.54
Netherlands, 14, 29, 40
Netherlands East Indies,14, 28, 59, 62,
93, 101, 115, 124
Neveling, C.H., 147
New South Wales, 23
New Zealand, 23
Ngali, 56
Nile valley, 26
Nooitgedacht Horse Breeders' Society,
148
Nooitgedacht pony, 147-148
Northern Territory, 31
nutritional needs of horses, 12,
101-102, 109, 196 n.56, 209 n.41.
See also pasture

Okluang Chula, 73
Olivier, Johannes, 53
Oman, 27
Orange Free State, 147
'other'
animals as, 6-10, 161 n.53
horses as, 153
oxen, 22, 26, 29, 124, 133, 1399, 207
n.

pack horses, 60, 66, 86, 87, 90, 106,
192 n.15. *See also* draught horses;
transportation and haulage, use of
horses for
Padang, 37
Paie, 56
Pajajaran, 38
Palatine Law of Siam, 69
Palembang. *See* Sumatra
Pampanga, 91, 92, 93, 94, 96, 99, 110
Pariaman, 35, 37
Pasay Country Club, 92
pasture, 12, 209 n.41

Indonesia, 35, 36, 37-38, 40, 42,
45, 47, 48, 61, 62, 63
Philippines, 100, 102, 117
South Africa, 24, 25
See also nutritional needs of
horses
Pasuruan, Residency of, 42, 59
Payakumbuh horse, 37
pedigree, 2, 11, 12, 13
Pegu, 29, 35
'Pegu Ponies', 29
Pekat, 52, 53
Penang, 29
periodisation, 8
Persia, 23, 34, 39, 69-70, 130
Persian Gulf, 23, 24, 27, 31
Persian horses, 34, 35, 36, 38, 40, 44,
47, 65, 71
and Cape Horse, 129, 131, 132
gifts to Siamese rulers, 72, 76-77,
78-79
influence on Javanese horses, 67
Peruvian Paso, 131
pet-keeping relations, 157, n.17
Phetracha, King 65, 70, 72, 74-78, 79
Philippine-American War, 95, 194
n.32
Philippine Horse (breed), 85, 100-101,
103, 105, 191 n.1, 197 n.2
and Spanish identity construction,
116-119, 121
size, 105-107, 108, 111, 113,
114, 117, 119, 120, 121, 198-199
nn.13-16
transformed by environment, 114,
117, 120
Philippine Racing Club, 193 n.23
Philippines
breeding, 108-111, 112-116, 119,
120, 199-202
climate, 114, 117-118, 204 n.75
colonial legacy, 100-103
deaths of horses, 95-100

dietary regime of horses, 101-102, 109, 196 n.56

environmental impact of horses 100, 101-103

export of equids, 21, 29, 76

feral horses, 95, 101, 102-103

floods, 95

health of horses, 92-95, 195 n.47

horse/people ratio, 101

horse population, 91, 93, 95, 96, 100-101, 102, 195-196 nn.52, 54

horses in colonial life, 86-92

Manila as mart for sea-borne horses, 31-32

racing in, 91-92, 100, 193 n.23

Spanish introduction of horses, 13, 85, 105

Philo, Chris, 2, 5

Phra Phutthabat shrine, 71

pigs, guilt connected to use of, 159 n.35

Pires, Tomé, 35, 37, 38, 42, 43, 54, 65

Plate, River, 23

Poja, 56

polo, 32, 69

polygenism, 117, 204 n.68

ponies, 14-15, 21, 28, 29, 31, 32, 68, 69

definition of, 163 n.84

See also Java ponies; 'South East Asia Ponies'

popular culture, animals in, 2

population density, impact on breeding, 37-38, 40, 41, 45-46, 47, 48

Portugal, 29, 34

postal service, use of horses in, 44

post-modern movement, 4, 159-160 n.41

power, horses as symbol of, 127, 128, 129, 139, 209, nn.37, 209

Prasatthong, King, 69, 72

Priangan horses, 39-40, 42, 43, 44, 47

Priangan upland valley, 39, 43, 45-46, 47, 48, 66

processions. *See* royal processions

quaggas, 126, 208 n.28

quarantine, 99

Queensland, 32

Quiapo, 90, 91

Raba, 61

'race', concept of, 147, 217 n.30

racehorses, 11, 24, 28, 32, 162 n.71

racing, 22, 69, 86, 162, n.71

and public *personae* of horses, 134

Cape, 131-132, 134-138, 139

Lesotho, 144

Philippines, 91-92, 100, 193 n.23

Southern Africa, 14, 15

Raffles, Thomas Stamford, 45, 178 n.13

railway, horse-drawn, 89

Rangoon, 29

Reenen, Daniel van, 132

Reinwardt, Caspar, 58, 182 n.53

Reniers, Carel, 72

Réunion (Bourbon), 21, 27, 58

Rhodesias, 26

riding, 22, 38, 43, 54, 57, 70-71, 118, 125, 132-133. *See also* saddle horses

Riebeeck, Jan van, 124-125, 127, 129, 131, 206 n.4, 208 n.22

Riouw, 58

Ritvo, Harriet, 2, 3, 4, 5, 6, 10, 11

Romanes, George, 7

Rose, Cowper, 136

Roti, 44, 45

royal courts

Java, 38-39, 41, 42, 43, 78

link between breeding and, 13, 36, 37, 39, 40, 41, 42, 46, 47, 48, 56, 172, n.17

Minangkabau dynasty, 37

Moghul, 43
Sumatra, 36–37, 172, n.17
Sumbawa, 54–56, 63
  See also Ayutthaya, court of
royal horse guards, 68, 72
royal processions, 13, 68, 71
rulers, gifts of horses to, 33, 34
ruwe, 56
Ryukyu Islands, 23

saddle horses, 87, 89, 90, 96,
    106–107, 113, 119, 192 n.15. *See*
    also riding
Said, Edward, 6
Saie, 56
Saint-Hilaire, Isidore Geoffroy, 111
San (Bushman; Khoisan), 124, 125,
    127, 128, 129, 142, 143, 206–207
    n.6
San Pascual, 91
Sanchez Mira, José, 106, 108,
    110–111, 119, 199 n.22
sandalwood horses, 46, 60
Sandalwood Island. *See* Sumba
Sangeang (Gunung Api), 53, 54, 55,
    56, 101
Sanggar, 52, 58
Savu, 44, 45, 60
Schouten, Joost, 68, 71
Semarang, 44, 59, 78, 80
Sembrani horses, 38
Sendoro, 41
Sewell, Anna, *Black Beauty*, 134
Seychelles, 27
Shepard, Paul, 4
ships, 15, 22, 24, 28, 45
    Arab, 30
    British, 30
    cattle boats, 59
    Chinese, 75, 78
    Dutch, 29, 30
    French, 28
    Javanese, 77

monsoon winds and use of, 58
perahu, 59
sailing ships, 25, 31
Siamese, 72, 73–74, 75
Spanish, 31
steamers, 25, 30, 59, 110
VOC ships, 34, 72, 73–74, 75–76,
    77, 79, 80, 134
Siam, 13, 34, 39, 44–45
    significance of horses in, 68–71
    See also Ayutthaya, court of;
    Thailand
'simian orientalism', 8
Singapore, 29, 58, 110
Singapore Chronicle, 53
Sirih Puan, 55
size of horses
    large horses valued, 30, 33–34, 39,
    42
    Philippines, 105–107, 108, 111,
    113, 114, 117, 119, 120, 121,
    198–199 nn.13–16
    small, Arab, 24, 31
    small, Southeast Asia, 13, 14, 33,
    106, 120
social analysis, categories of, 8
social change, theories of, 8
Somerset, Lord Charles, 24, 136,
    137–138
Songkhla, 69
Sotho, 23, 142, 143, 150, 210 n.52.
    See also BaSotho people
South Africa, 9
    breeding, 147–148, 149
    committee to investigate
    indigenous breeds, 147
    export of equids, 21, 22, 23, 24–26,
    28
    export of horses overland to
    African countries, 21, 26
    horse culture, 128
    'mineral revolution', and export
    price of horses, 25, 28

See also Cape, South Africa;
Southern Africa
South African Stud Book, 148
South African Turf Club, 137
South African War, Second, 26, 93, 144
South America, 28, 93, 131
South China Sea, 22, 74, 110, 149
'South East Asia Ponies', 14-15, 124, 127
South Sulawesi, 30
Southeast Asia
breeds of horses, 12-13
diseases, 21, 103
export of equids, 21-22, 29
impact of colonisation on horses, 13
internal markets in, 28-32
large horses valued, 30, 33-34, 39, 42
population of horses, 23
small size of horses, 13, 33, 106, 120
stud stock from, 132
trade contacts, 67
See also specific countries
Southern Africa, 1, 13
breeds of horses, 14-15
diseases, 21, 22, 25
European colonisation and settlement 1, 123
export of equids, 21-22, 23, 24-26
horse as 'personality', 134
imports of equids, 14, 23
internal demand for horses, 26
population of equids, 23
South West, 26
See also Cape, South Africa;
Lesotho; South Africa
Southwest Asia, 52
Spain
administration of Philippines, 107, 108, 110, 113, 114, 115-121

breeding programmes, 108
connections with Macau, 29
conquest of Americas, 1, 9
defeat in Spanish-American War, 31, 32
introduction of horses to
Philippines, 13, 85, 105, 1113
reintroduction of horses in
Americas, 32, 94, 131, 155 n.3
Sparrman, Anders, 126
stables, 56, 59, 70, 79, 90
stallions, 34, 47, 150
Cape, 129, 133-134
castration 59, 115-116, 203 n.60
Java, 59, 73, 79
Philippines, 17, 91, 107, 108, 109, 110, 112, 113, 115, 116, 120, 200-201 n.33
Sumbawa, 54, 56, 59
steam power, reduces need for
horses, 48
Stel, Simon van der, 129
stud farms
Bima, 56
Cape, 133
Lesotho, 144, 145, 216 n.25
Melck stud, 213 n.92
off Ceylon coast, 34, 47, 57, 79, 171 n.1
Philippines, 85, 109
South Africa, 147-148
Sua, King, 70, 78
Sudan, 28
Suez Canal, 25
Sulawesi (Celebes), 34, 47, 55, 57, 60
Sulu Horse, 197 n.2
Sulu Islands, 13, 85, 91
Sumatra, 34, 52
breeding, 35-38, 46, 47, 48
breeds of horses, 66, 172 n.13
eating of horse flesh, 194-195 n.40
export of horses, 21, 22, 29, 44, 65, 85

imports of horses, 39, 72
stud stock from, 34
See also specific placenames
'Sumatran horse', 65, 172-173 n.18
Sumba, 30, 31, 44, 45, 46, 60, 66, 67
Sumbawa, 46, 51-64, 171 n.2
   breeding/breeds, 45, 46, 53-54,
     56, 59-60, 60, 61, 106, 183 n.64
   colours of horses, 54
   exchange of horses for gifts and
     weapons, 57, 63, 205 n.79
   export of horses, 14, 30, 43, 44,
     45, 57, 58-59, 60, 63, 64, 127, 209
     n.35
   horse-keeping methods and
     environment, 60-63, 101
   horse/people ratio, 62
   imports of horses, 65
   introduction of horses, 52-54
   royal horses at courts of, 54-56
   seasonality of horse trade 58
   topographical names, 176-177
   value of horses in, 56-60, 63, 64
   veterinary service, 59, 185 n.82
   Western, 53, 54, 56, 177 n.5
Sumbawa Besar, 59, 60, 64
Sumbing, 41
Sumedang, 39
Sunda, 38
Surabaya, 42, 43, 44, 57
Surakarta, 41
surra (trypanosomiasis), 21, 27, 32,
   62, 63, 92-95, 98, 123
Swart, Sandra, 1-18, 123-139,
   141-150, 153-154

Tachard, Guy, 74
Tagalog region, Northern, 93, 94, 96,
   99
Tagalog region, Southern, 87, 93, 94,
   96, 99
tails, cutting of, 70, 136, 138
Taiwan, 34

Taliwang, 59
Tambora (sultanate), 45, 52, 53, 58
Tambora horses, 45, 53-54
Tambora, Mount, eruption of, 30,
   53-54, 58
Tamil traders, 29
Tant, Gideon, 76
Tegal, 44
Thailand. *See* Siam
Thaisa, King 65, 78-80
theft of horses, 96-97, 129, 143
Thirsk, Joan, 12
Thompson, George, 132
Thorn, William, 53
Thorndike, Edward Lee, 7
Thornton, R.W., 145
thoroughbred horses, 11-12, 144,
   145, 162-163 n.74
   English, 12, 15, 108, 131, 132, 133,
     134-135, 137, 138, 139, 213 n.90
Thunberg, Charles, 133
Tibet, 29, 32
Timor, 29, 30, 44, 45, 46, 60, 66
Timor pony, 66-67
tobacco cultivation, 41
Tou Juran, 56
Tou Kamutar, 56
tournaments. *See* jousting
   tournaments
trade in horses
   maritime, Indian Ocean, 21-32,
     149
   seasonality of, 58
   strategic nature of, 22-24
   See also Arab traders; Dutch
     United East Indies Company; and
     specific countries and regions
trade networks, 1-2, 13
tramcars, horse-drawn, 90
tramway, mule-drawn, 89
transport and haulage, use of equids
   for, 26, 32, 149

transport and haulage, use of horses
   for, 22, 32, 48, 149
   Cape, 124, 129
   Indonesia, 37, 40
   Java, 29, 66
   Lesotho, 141
   Philippines, 86-91, 110, 192 n.15
   South Africa, 26
   Sumbawa, 56-57, 60, 63, 64
Tripitaka 65
triple (*trippel, tripple*) 131, 132, 133,
   211 nn.62, 73, 76
tropicality, 111, 121
tropics, European beliefs about,
   116-118, 120-121
trotting, 131, 132, 133, 211 n.62, 212
   n.72
Trunajaya War, 40-41
trypanosomiasis (surra), 21, 27, 32,
   62, 63, 92-95, 98, 123
tsetse fly, 21
Tuban, 42, 43
Tuharlanti, 55

U'a Pua, 55
ungulate irruption, 9-10
United States, 5, 9, 26, 32, 107
   administration of Philippines, 90,
   94, 98-100, 102, 107, 114, 116
   export of equids, 26, 32
Uruguay, 28

valuation of horses, 70-71
Van Braam Morris, D.F., 55, 56
Van Sittert, L. 9
Vanrenen family, 26
Vereenigde Oost-Indisch Compagnie
   (VOC). *See* Dutch East India
   Company
veterinary services
   Philippines, 98-100, 109, 195 n.47
   Sumbawa, 59, 185 n.82
Vietnam, 34

Visayas, 93, 94, 99, 101
Vliet, Jeremias van, 68, 69, 71

Walers, 23, 28, 29, 31, 32
warfare. *See* military use of equids
Wera, 56
Western Australia, 28, 31
Wilkes, Charles, 91
Willis, Roy, 4
Witjens, J.C. 52
Wolch, Jennifer, 3, 4, 128
Wolf, Eric, 153
Woltemade, Wolraad, 134, 213 n.85
women
   devaluation of, 8, 160 n.50
   history of, and revision of
      historiography, 8
   Lesotho, 215 n.4
World War I, 26, 31, 32

Yogyakarta, 41
Young, Sir George, 136
Yunnan, 12, 29, 32

zebras, 26, 123, 126, 208 nn. 30-31
Zollinger, Heinrich, 54, 55, 58, 179
   n.19
zoos, 5, 9
zootechny, 111, 112, 113, 119, 120,
   121
Zulu, 143